W9-BGG-113

NEW ENGLAND WHITE WATER RIVER GUIDE

by Ray Gabler

Appalachian Mountain Club

*This book is
dedicated to my wife
Rose Mary*

New England White Water River Guide

Copyright 1981 © by Ray Gabler

Second Edition
  first printing   April 1981
  second printing   August 1983

Acknowledgments

I would like to thank the following people who have helped me in preparing this book: Ted Acton, Dave Cooper, Barbara Cushwa, Bill Cushwa, Renée LaFontaine, Rose Mary Gabler, and the U.S. Geological Survey Water Resources Division (Boston). Without their help, this effort would have been much harder and not nearly so pleasant.

I would also like to thank the following people for their work on the book: Les Fry, hand signal drawings; Debbie Arnold, cover photo; Dave Cooper, maps; Robert Saunders, design; Sally Carrel, editing; Michael Cirone, production, and Arlyn Powell, publisher.

ISBN 0-910146-33-0

Contents

ABOUT THE A.M.C.

The Appalachian Mountain Club is a non-profit volunteer organization of over 25,000 members. Centered in the northeastern United States with headquarters in Boston, its membership is worldwide. The A.M.C. was founded in 1876, making it the oldest and largest organization of its kind in America. Its existence has been committed to conserving, developing, and managing dispersed outdoor recreational opportunities for the public in the Northeast and its efforts in the past have endowed it with a significant public trust; its volunteers and staff today maintain that tradition.

Ten regional chapters from Maine to Pennsylvania, some sixty committees, and hundreds of volunteers supported by a dedicated professional staff join in administering the Club's wide-ranging programs. Besides volunteer organized and led expeditions, these include research, backcountry management, trail and shelter construction and maintenance, conservation, and outdoor education. The Club operates a unique system of eight alpine huts in the White Mountains, a base camp and public information center at Pinkham Notch, New Hampshire, a new public service facility in the Catskill Mountains of New York, five full-service camps, four self-service camps, and nine campgrounds, all open to the public. Its Boston headquarters houses not only a public information center but also the largest mountaineering library and research facility in the U.S. The Club also conducts leadership workshops, mountain search and rescue, and a youth opportunity program for disadvantaged urban young people. The A.M.C. publishes guidebooks, maps, and America's oldest mountaineering journal, *Appalachia.*

We invite you to join and share in the benefits of membership. Membership brings a subscription to the monthly bulletin; *Appalachia;* discounts on publications and at the huts and camps managed by the Club; notices of trips and programs; and, association with chapters and their meetings and activities. Most important, membership offers the opportunity to support and share in the major public service efforts of the Club.

Membership is open to the general public upon completion of an application form and payment of an initiation fee and annual dues. Information on membership as well as the names and addresses of the secretaries of local chapters may be obtained by writing to: The Appalachian Mountain Club, 5 Joy Street, Boston, Massachusetts 02108, or by calling during business hours, 617-523-0636.

CAUTION

It should be realized at the outset that white water canoeing can be a dangerous sport. This book is not a guarantee to safe or enjoyable boating. The success or enjoyment of a trip depends primarily on you, the boater, in your own paddling skills, your ability to cope with unexpected problems, and in your judgment. This book should only be used as a guide. Rivers and their surroundings will change with time, so don't take the descriptions presented here as gospel. Make observations on your own prior to running.

Introduction

This is a book describing the white water rivers of New England. It is an effort to better acquaint the paddler with some of the more frequently traveled waterways and to capture the interest and imagination of the novice in as exciting and challenging a sport as there is. This book is not intended to be an instructional manual on the techniques of canoeing or kayaking,* although the explanatory parts are written so that a neophyte may also gain a good working knowledge of the factors to be considered when contemplating a river trip. This, then, is a guide to the rivers themselves; it is an atlas of aquatic highways that are as yet unpaved, free flowing, and exciting.

The river trips described in this book are designed to take less than a day to complete. Most true white water runs in New England are fairly short; hence, they can frequently be repeated several times in a day. Camping out along the trip is not necessary and this cuts down on the duffel needed to be carried in the boat. Many closed boats don't have room for anything except the paddler in any case. No week-long wilderness excursions are discussed.

The rivers considered here are those that appear most often on the schedules of the various New England boating clubs, but this guide is by no means a comprehensive survey of *every* river section that these clubs have ever run. This book is written to aid boaters in their knowledge of the river and to help make the journey a much safer one so that many more trips will be taken.

In preparing this guide, much use has been made of information that is readily available in our present-day technological society: gage readings, discharge measurements, river fluctuation data, and topographic maps. One may wonder how, without benefit of all this information, early-day canoeists first made it down those foreboding, unexplored rivers that local legend and tradition had built into overpowering ogres. Apparently what those paddlers lacked in knowledge, they made up in skill, cunning, and broken boats. Since not everybody possesses such qualities or wishes to pursue such grandiose ventures, these guides are necessary: the more a prospective voyageur knows about an unfamiliar river before he sets a boat on it, the better his chances of having a safe, enjoyable time. It is with this philosophy that the following is presented.

* See the Appendix for a list of instructional manuals.

Photo — Dave Cooper

White Water Canoeing in New England

Along with its heritage of early America, its folklore, and its sectionalism, New England possesses a topography that is not only beautiful to look at, but is also ideal for harboring white water rivers. Tucked into narrow folds between rapidly rising peaks, these aquatic arteries drain the snow-capped hills during the spring runoff. Like the states that comprise the area, most canoeable rivers here are small, with roads running alongside and towns nearby. Very few rivers are completely isolated from civilization, and, as more and more people move northeastward, this situation will worsen. Although wooded hills generally enshroud most rivers, their valleys are not extremely deep. Where there are no hills there is no white water.

Rocks of all shapes and sizes litter these narrow watery avenues, and it is these very rocks that give the rivers their character, and the paddlers their challenge. New England rivers are distinctive not only for their size, but also for the preponderance of stones, rocks, and boulders that obstruct and shape the downward-sloping channels. Of all the traits that characterize rivers in the Northeast, the abundant supply of rocks in the riverbed is the most outstanding.

The white water season in New England starts when the ice breaks and the rivers start to flow free and clear. Most serious paddlers begin the season in early March: when the maple syrup flows, so do the rivers. Both weather and water are very cold at this time, so it is strongly recommended that canoeists wear wetsuits. A long swim in cold water without a wetsuit can not only dampen your spirits, it can also present a real danger to your life. Snow cover that was used for cross-country skiing just a week or month ago now makes scouting rapids difficult; it can also block back roads making river access impossible. But even with all of these problems, early spring canoeing can still be enjoyable. It gets you out of the house, into your boat, and on the river. It uplifts your spirit after a long, sequestering winter. It signals the start of a new spring season, and prepares you for the better canoeing days ahead when the weather warms and the rivers roar.

1

The rivers that open up first are found in southern Connecticut; paddling usually starts there early in March. The Salmon and the Shepaug are the two most popular runs since they are reasonably easy and can be run in both open and closed boats for several weekends. The Ten Mile-Housatonic trip isn't quite so popular, but it has the same characteristics. From here, boaters move northward with the thawing spring, hoping to catch each run with just the right amount of water for an exciting trip.

In Massachusetts, the Quaboag is one of the first rivers to rid itself of ice. It also enjoys a relatively long season since it is lake-fed and not so subject to oscillations of the weather as are other rivers. The Westfield system offers a wide variety of trips (seven) and difficulties (Class 1-4). The Farmington above New Boston gives one of the more challenging early season tests to the boater, as the water stampedes downward among a plethora of rocks and tight channels. The Millers in northern Massachusetts has two sections that can be playful; the lower run passes along Route 2 where rubber-necked drivers contort to keep the boater's river antics in view.

For all of its hills, Vermont has surprisingly few really difficult runs. Most rivers here are only moderately difficult, although a few will give cause for concern and require expertise in high water. In general, Vermont rivers are best suited for open boats. Vermont is the state with the most rivers whose names start with the letter W; it also has the fewest number of billboards. The West is Vermont's most famous canoeing river, since many national championships have been held on one of its rapids known as the Dumplings. The West also has a dam which can generate water when other rivers nearby are dry. In addition, Vermont can boast of a river whose name has thirteen letters, four of which are O, and a bridge that crosses another river where the casual pedestrian can spit from a height of over a hundred and sixty feet. But the single most impressive feature of Vermont rivers is the countryside surrounding them. From pleasant hills enshrouding the Green, the White, and the Williams to the rugged outback around the Winhall and the Wardsboro, Vermont displays a panorama of distinctive scenery that adds to the boater's pleasure.

If the Vermont rivers are pleasant and moderate, then those in New Hampshire are mean and arrogant. In New Hampshire, you'll meet such rapids as Surprise, Gum Drop, Egg Beater, The Staircase, and Freight Train. None of these are for the faint-hearted. In southern New Hampshire, the Contoocook will offer a most violent and turbulent ride in high water, while the nearby Souhegan is calmer and more suited for open boats. The Blackwater near Concord races down a course between slalom gates made of summer homes and ends in serpentine rapids where you'll have to store your false teeth in a pocket for fear of losing them. The Piscataquog is serene and gentle, generally being a favorite of beginners and those just switching from lake to river canoeing. The White Mountain rivers as a group are rated as the most interesting in New Hampshire. The Swift and the Pemigewasset are particularly noteworthy. The Swift is narrow and fast, while the Pemi is large, formidable, and will run over you like a tank in high gear. Both deserve respect. The Ellis, the Mad, and the Smith are small in size, but the labyrinthine course of each necessitates a strong sense of boat control and maneuverability. The Saco can be either playfully frisky or devilishly deceptive depending on water level, but in any case it flows through the most spectacular valley in the White Mountain area.

In Maine, the Rapid and the Dead are impressive for their scenery; the Rapid is also noteworthy for its difficulty. Both rivers are hidden away from the populous parts of New England and they reflect the rugged Maine territory in their waters. The Dead is most suited for open boats while the Rapid will deal sternly with them. The three representatives from New York are the Sacandaga, the Boreas, and the Hudson. The Boreas is small, but it descends from the hills like a herd of stampeding bulls. The Hudson is one of the largest rivers described in this book, an angry giant in the spring. It is not unusual to see *pieces* of aluminum canoes littering its banks.

For those who would like to know what the toughest rivers are, the following will present an impressive list: Boreas, Contoocook, Hudson, Pemigewasset, Rapid, and Swift. For rivers heavy on scenery consider these: Boreas, Dead, Green, Hudson, Rapid, and White. Good open boat rivers include: Green, White, Piscataquog, Suncook, Westfield, Shepaug, and Salmon.

When the sun starts to get hot and temperatures reach into the seventies and eighties, the main canoeing season is over. This usually happens during the last part of May or the beginning of June. Rain is a big factor then because it can help a crippled season limp along for another weekend. But once the sustained snow melt is over and the last batch of white water eggs is hatched, boaters put away their paddles. Summertime is not all that bleak, however, since there are several rivers that are dam-controlled and run throughout the hot season. They are not always that exciting, but they are there. The Tariffville section of the Farmington in Massachusetts has a 1.5 mile stretch that runs Class 2, good for both open and closed boat practice. The distance may not sound like much, but the Farmington packs a lot of variety in that short space. The Androscoggin in northern New Hampshire has a long section that is mostly flat, although rapids of Class 1-2 can be found. In Maine, Flagstaff Dam on the Dead occasionally releases water for a weekend frolic. And, if you don't mind taking the trouble to get there, the Rapid offers a blend of beauty and excitement that is hard to match. If the rivers don't interest you there is always the tidal rip at Cohasset where a narrow inlet forms a Class 2-3 standing wave-hole combination that's great for playing and practicing. Ocean kayak and canoe surfing have also become popular lately, where the standing waves move and you can do an unexpected somersault when your bow catches in the sand. Add boat patching, story swapping, cycling, and canoe polo to this list, and you have a fair idea of what white water boaters do in the summer.

Fall canoeing usually picks up around September and October, relying on water releases on the Dead, West, and Farmington Rivers. These releases are coordinated to overlap weekends, so boaters can take full advantage of a reservoir lowering procedure. During the rainy season in late October and November, some rivers hold enough water for enjoyable but not strenuous sport. Those in this category include the Lower Ashuelot, Lower Millers, Contoocook, and Quaboag. The weather in the fall is similar to that in the spring although typically the water is warmer. The fall season is also the most aesthetically pleasing, since the New England hills blaze in a vast panorama of color. Brilliant scarlets, yellows, and oranges paint every river valley, interrupted only by an occasional evergreen stroke of the artist's brush. Combine this with white water dancing in front of you, crisp air, and reflected sunlight glittering off the water, and you can easily understand why fall is a truly spectacular time of the year for boating.

Winter canoeing is not too popular because of the ice cover and cold temperatures. However, winter is a lovely time with its snow-covered trees and picturesque ice formations. An impromptu gambol down a small, easy stream on a windless, sunny day is a joy that can bring a zest and exhilaration that are quite different from battling titanic waves and cavernous souse holes.

So, depending on your desires, New England can offer much in the way of variety, not only in its rivers but also in its scenery and its seasons. If you want rugged weather and water, it's here, and so are a pleasant summertime sun and a lazy paddle down a lazy stream.

Organization

This guide is essentially organized in two parts. The first consists of definitions and descriptions of terms that are used throughout the rest of the book. This first section is used to set the stage for the river descriptions themselves. It is the place where explanations are given for precise meanings of particular words or conventions. Degree of difficulty, river level, average gradient, maximum gradient, and the use of gages are among the topics presented. Without a clear understanding of these concepts, the river descriptions later in the book cannot be fully utilized. Discussions on safety, lifejackets, hypothermia, and scouting, along with river hand signals, are also included. Although this book is not intended to be a teaching manual, the first section is somewhat instructional. The first part also includes a chart that condenses the key characteristics of every river, so the reader may have easy access to a plethora of information. The chart is also useful for comparisons.

The second part of the book is the river descriptions themselves; included are explanations of starting and stopping points, specific difficulties to be found, gage locations, and a paragraph or two giving an overall feeling for a river's character. A map accompanies each river trip so the reader has a pictorial as well as a verbal representation. Rivers are listed alphabetically; if there is more than one trip per river, trips are labelled A, B, C, with the "A" trip being the closest to the headwaters. These descriptions are usually based on notes taken while the river was actually being paddled, although a few writeups are a result of a scouting trip only. Descriptions are sequential and are presented just as a paddler would encounter the various rapids and obstacles. *The directions left and right are always with respect to a boater facing downstream.*

The descriptions are as accurate as possible, although the reader should be aware that nature changes with time, and that these accounts should not be taken as being absolute. Where deviations from the descriptions are found, use your own skills and judgment. No guidebook can substitute for these.

Photo — Lynn Williams

Part One

River (State)	Trip	Difficulty	Boston (miles)	Springfield (miles)	In	Out	Trip Distance Miles	Avg. Gradient feet/mile
Ammonoosuc (N.H.)	A	2-3	160	200	River Bend	Pierce Bridge	3	43
	B	1-4	160	195	Pierce Bridge	Route 116	7	35
Androscoggin (N.H.)		1-2	225	260	Errol	Pontook School	20	4
Assabet (Mass.)		1-2	15	95	Route 117	Route 62	1.5	10
Ashuelot (N.H.)	A	3-4	100	90	Marlow	Gilsum Gorge	3.5	63
	B	2	95	85	Gilsum Gorge	Shaw's Corner	4	30
	C	3-4	85	60	Ashuelot	Hindsdale	3.5	52
South Branch Ashuelot (N.H.)		3-4	75	80	Troy	Route 12	2.5	80
Bantam (Conn.)		1-2	130	60	Stoddard Rd.	Shepaug River	5.4	25
Bearcamp (N.H.)		2-4	130	195	Bennett Corner	Whittier	3.5	32
Black (Vt.)		2+	165	115	Whitesville	Perkinsville	7.5	27
Blackledge (Conn.)		2	105	50	Route 66	Salmon River	5.9	20
Blackwater (N.H.)		1-4	85	130	Route 127	Snyder's Mill	2.5	24
Boreas (N.Y.)		4-5	222	142	Route 28N	Minerva Bridge	7	45
Branch-Pascoag (R.I.)		1-2	60	70	Harrisville	Glendale	4.8	8
Chickley (Mass.)		2-3	108	65	Route 8A	Deerfield River	4.4	77
Cold (N.H.)	A	2	110	95	South Acworth	Vilas Pool	5.5	43
	B	2	115	90	Alsted	Drewsville	2.0	40
Contoocook (N.H.)	A	2	65	80	Jaffrey	Peterborough	5	37
	B	3-4	90	105	Hillsboro	Henniker	6.2	23
Dead (Maine)		2-3	270	350	Spencer Stream	West Forks	15	28
Deerfield (Mass.)		2	120	65	Bear Swamp	Route 2	9.5	22
Ellis (N.H.)		4	165	230	Route 16	Harvard Cabin	3.0	87
Farmington (Mass.-Conn.)	A	2	110	25	Otis Bridge	Route 8	2.4	27
	B	3-4	110	30	Route 8	New Boston	3.0	75
	C	1-2			Riverton	Route 44	12.4	14
	D	2-3	100	25	Route 189	Route 187	1.5	15

River Levels

Max. Gradient feet/mile	Shuttle (miles)	Scenery	Too Low	Low	Medium	High	Too High	Flood	Gage Location	Page
50	3	Good		3.6		5.1		10	Bath Bethlehem	57
				3.6	5.0C	5.0				
55	7	Fair			5.0C	5.0S		10	Bath Bethlehem	61
					4.5C	4.5C				
14	20	Good		1500 CFS					Errol	65
	1.5	Unfortunate	2.5	3.0					Maynard	69
80	3.5	Good	3.6	4.1	5.5C	7.5C			Gilsum Gorge	72
50	4	Good		4.6	5.7	6.5	7.5		Gilsum Gorge	77
80	3.5	Poor			4.8C	5.3C		13	Hindsdale	80
100+	2.5	Fair	0.0						Route 12	84
32	5	Good		0.5					Rt. 47 Bridge	88
	3.5	Good	0.5	1.0					Whittier	91
40	7.5	Fair	0.5	1.5					Covered Bridge	94
30	5	Good	1.4	3.0					Comstock Bridge	97
50	2.5	Fair		3.4C	4.3C				Webster	100
100	11	Excellent	4.8C		5.5C	7.0C	7.0C		North Creek	104
90	3.5		2.9						Forestdale	109
100	4.0	Good	4.3		5.3C	5.3S	5.3		Shattuckville	112
53	5.5	Good	3.4		5.2	8.0			Drewsville	116
40	2.0	Good	3.4		5.2	8.0			Drewsville	119
100	5	Good		2.8					Peterborough	122
60	6	Good	5.6	7.4C	9.0C	10C	10C	10	Henniker	125
50	20	Excellent		800 CFS	1300 CFS				Flagstaff Dam	130
40		Good	2.6	3.9					Charlemont	135
100+	3.0	Good	1.3C	2.1C	2.9C				Route 16	139
50	2.4	Good		4.0	5.0				New Boston	143
100	3.0	Fair	3.6	4.0C	4.5C				New Boston	146
		Fair								151
40	2	Fair		1.9	3.2C				Tariffville	154

River (State)	Trip	Difficulty	Boston (Miles)	Springfield (Miles)	In	Out	Trip Distance Miles	Avg. Gradient feet/mile
Gale (N.H.)		1-4	155	135	Franconia	Ammonoosuc River	7.5	25
Green (Vt.-Mass.)	A	2-3	110	65	Green River	West Leyden	6.8	30
	B	2			West Leyden	Covered Bridge	5.6	31
Housatonic (Conn.)		2-4	155	85	Ten Mile River	Gaylordsville	2.2	16
Hudson (N.Y.)		4	220	140	Indian River	Route 28	12.5	29
Indian (N.Y.)		3-4	220	140	Lake Abanakee	Hudson River	1.5	
Jeremy (Conn.)		2	105	50	Old Route 2	Salmon River	2.8	
Mad (N.H.)		3-4	135	185	Waterville Valley	Goose Hollow	8.5	85
Mascoma (N.H.)		2-3	125	130	Mascoma Lake	Lebanon	4.0	38
Millers (Mass.)	A	2-3	65	75	South Royalston	Athol	7.0	32
	B	2-3	70	50	Erving	Millers Falls	6.5	29
North (Mass.)		2	110	55	Halifax Gorge	Colrain	7.0	31
Ompompan-oosuc (Vt.)		2-3	195	145	Strafford	Rice Mills	9	45
Otter Brook (N.H.)		3-4	85	90	East Sullivan	Otter Brook Park	3.3	80
Pemigewasset (N.H.)	A	4	140	190	Kancamagus Hw.	Loon Mt.	2.5	72
	B	4	135	185	Loon Mt.	Route 93	3.4	65
South Branch Piscataquog (N.H.)		1-2	70	115	New Boston	Goffstown	8	13
Quaboag (Mass.)		2-4	80	35	Warren	Route 67	5.5	31
Rapid (Maine)		4	235	270	Middle Dam	Cedar Stump	4.5	40
Sacandaga (N.Y.)	A	2-3+	195	115	White House	Campground	8.5	36
	B	3	195	115	Route 8	Route 30	2.7	43
	C	3-4	195	115	Route 8	Route 30	2.0	55
Saco (N.H.)		3-4	170	225	Crawford Notch	Bartlett	6.5	40
Salmon (Conn.)		2	110	55	Jeremy River	Route 16	3	27
Sandy (Conn.)		3-4	125	45	Sandy Brook Road	Route 8 Bridge	3	80

Max. Gradient feet/mile	Shuttle (miles)	Scenery	River Levels						Gage Location	Page
			Too Low	Low	Medium	High	Too High	Flood		
80	5.5	Excellent	.5		1.0 1.0C				Franconia	158
50	6.5	Excel.	2.8	3.9					East Colrain	162
40	6.0	Good		3.9					East Colrain	166
16	2.2	Fair				7.1C		8	Gaylordsville	282
80	14	Excellent	2.8		5.5C	6.3C	7.0C	10	North Creek	168
		Good								168
		Good		1.4	3.0				Comstock Bridge	235
100+	8.5	Excellent		1.2C	2.5C				6 Mile Bridge	173
80	3.5	Fair			2.3C	2.3			Route 4A	176
60		Fair	4.4	6.0	7.0				S. Royalston	179
64	6	Fair	2.9		4.4/ 4.8C			10	Farley	183
50	7.0	Good	3.3	4.6	5.3				Shattuckville	187
60	9	Good								190
100	3.3	Fair		1.5C	1.5	3.0C			East Sullivan	193
80	2.5	Good	0.2	1.0C	1.5C	2.0C			Kancamagus Br.	197
100	3.4	Fair		1.0C	1.5C	2.0C			Kancamagus Br.	203
30	8	Good		315	318 6.8				Route 13 Grasmere	206
85	6.0	Poor		3.9	4.4 5.5C	5.5	6.0		West Brimfield	209
80	4.5	Excellent			1200 CFS-C				Middle Dam	215
70	12	Good		3.3	5.0C	5.0C	5.0	6.5	Hope	221
				3.1	5.5C	5.5S	5.5	7.9	Griffin	221
50	3	Good	3.3		5.5C			6.5	Hope	224
			3.1		6.4C			7.9	Griffin	224
80	2	Good	3.3		5.0C	5.5C		6.5	Hope	227
			3.1		5.5C	6.4C		7.9	Griffin	227
60	6.5	Excellent	0.5	1.0C		4.0C			Bartlett	230
50		Good		1.4		3.0			Comstock Bridge	235
100	3	Good	2.1	2.6	3.3C		3.3		Rt. 8 Bridge	239

11

River (State)	Trip	Difficulty	Driving Distance Boston (Miles)	Driving Distance Springfield (Miles)	In	Out	Trip Distance Miles	Avg. Gradient feet/mile
Saxtons (Vt.)		2-3	135	90	Grafton	Saxtons River	7.5	46
Shepaug (Conn.)	A	2-3	135	65	Route 341	Route 47	7.0	40
	B	1-2	140	70	Route 47	Route 67	10.5	18
Smith (N.H.)		4	115	145	Route 104	Bristol	2.0	90
Souhegan (N.H.)		2-3	60	95	Greenville	Wilton	3.5	50
Stony Brook (N.H.)		3	60	110	Route 31	Wilton	2	65
Sugar (N.H.)		2-3	120	110	Newport	Route 103	2.5	34
Suncook (N.H.)		2-3	85	150	Pittsfield	N. Chichester	6	20
Swift (N.H.)	A	1-3	155	225	Bear Notch Road	Rocky Gorge	4.0	22
	B	3-4	150	220	Rocky Gorge	Gorge	3.5	40
	C	4	145	215	Gorge	Kancamagus Hw.	4.0	80
Ten Mile (N.Y.)		1-3	160	110	South Dover	Gaylordsville	4.2	12
Waits (Vt.)		2-3	205	160	Waits River	Route 25B	10	38
Wardsboro (Vt.)		3-4	135	95	Wardsboro	West River	4.5	
West	A	3	135	85	Ball Mt.	Salmon Hole	2.5	40
(Vt.)	B	2-2+	130	80	Salmon Hole	Route 100	3.5	30
North Branch Westfield (Mass.)	A	1-2	125	45	W. Cummington	Cummington	6.2	33
	B	2-3	120	35	Cummington	Chesterfield Gorge	7.2	36
	C	1-3	125	30	Chesterfield Gorge	Knightville Dam	9.2	12
	D	3	105	30	Knightville Dam	Huntington	5.2	17
Middle Branch Westfield (Mass.)		2-3	115	35	River Road	Littleville Dam	7.0	43
West Branch Westfield (Mass.)	A	3-4	110	30	Bancroft	Chester	6.0	50
	B	2-3	105	25	Chester	Huntington	7.5	30
White (Vt.)		1-2	205	160	Rochester	Gaysville	16	11
Wild Ammonoosuc (N.H.)		2-3	155	165	Stillwater	Route 302	2	90
Williams (Vt.)		2	140	95	N. Chester	Brockway Mills	7.5	16
Winhall (Vt.)		3-3+	140	90	Grahamville School	Londonderry Rd.	4.5	62

Max. Gradient feet/mile	Shuttle (miles)	Scenery	River Levels Too Low	Low	Medium	High	Too High	Flood	Gage Location	Page
60	7.5	Fair	4.2		5.2				Saxtons River	242
50	7.0	Good		0.5	1.0	2.0			Rt. 47 Bridge	245
50	8.5	Good	-0.4	0.0					Rt. 47 Bridge	248
100	2.0	Good	-0.5	0.5C		1.5C			Cass Mill Rd. Br.	251
65	4	Good	0.5	1.0	2.5C	2.5			Old Powerhouse	255
	2	Fair	0.0	0.5					Rte. 31 Bridge	259
40		Good	1.8		4.0			10	W. Claremont	262
30	8	Good	4.3				10	10	N. Chichester	265
	4.0	Excel.	2.4		3.1				Bear Notch Rd.	270
	3.5	Good	0.5	1.3C		3.0C			Gorge	273
100	4.0	Good	0.5	1.3C		3.0C			Gorge	277
15	6.0	Good				4.6 7.1			Webatuck Gaylordsville	282
50	10	Good	0.4		2.0				Rte. 25B Bridge	286
	4.5	Good	0.0						West River Confluence	290
50	2.5	Good			7.2C/ 6.2	7.2			Jamaica Park	294
40	3.5	Fair		6.2	7.4				Jamaica Park	299
50	6	Fair	0.0		2.0 4.0				Route 9 Bridge Huntington	302
60	5.4	Good	0.0 2.0	1.0	2.0 4.0				Route 9 Bridge Huntington	305
40	16	Good	0.0	1.0	2.0 4.0				Route 9 Bridge Huntington	309
45	5.5	Fair		4.5	5.0				Knightville Dam	312
50	7.0	Good	0.0	2.2 3.7	3.0				N. Chester Huntington	316
100	6.3	Excel.	2.2	3.0	4.0C				Huntington	319
40	7.5	Fair	2.0	3.0	4.0				Huntington	322
	16	Excellent		6.0				18	W. Hartford	326
100	2	Good	3.6	5.0					Bath	330
34	7.5	Fair	2.0	3.0					Brockway Mills	333
100	4.5	Fair	0.5	1.0	1.5 2.0C				Rte. 100 Bridge	336

River Characteristics

The terms described here are used in the chart on pages 8-13, and in the specific river descriptions in Part II.

RIVER

The rivers covered here are listed alphabetically. Names of the rivers are those in common usage and the ones found on Geological Survey maps. The state(s) in which the river is located is indicated in parentheses.

TRIP

If there is more than one trip reported for a given river, the trip closest to the headwaters is designated *A*; the next one downstream, *B*; etc. In cases where there are trips on different branches of the same river, the branches are listed separately and the trips are listed under the respective branches, e.g., see the Westfield. Where only one trip on a river is described, no designation is given in the Trip column.

DIFFICULTY

As one might expect, the difficulty of a particular trip is dependent upon many factors; discharge rate, gradient, complexity of channel, and kinds of obstacles are among the more obvious. Any one of these factors could dominate and make a tough trip tougher. It is necessary to consider them all. How, then, does one objectively make a quantitative rating of difficulty? The answer is, it's not easy. However, over the years an international system of grading rivers or individual rapids, on a scale from one to six, has evolved to be probably the most frequently encountered rating scale and it is the one used in this guide.[*] The difficulty is in getting people to agree with the description of each classification and then to agree on a rating for a particular river. Experts tend to give lower ratings to most rivers than the average paddler, whereas neophytes tend to overrate everything. Paddlers who have mastered a particular river also tend to underrate it while those who have had trouble on the same river overrate it. Thus, this honest difference of opinions makes it necessary, when accepting judgment of a river's difficulty from another boater, to also make a judgment on the other's paddling ability and experience. This is just as important as the classification itself.

Before discussing how the ratings were assigned, a description of the six classes will be presented. These descriptions are not by any means "official," but are a hybrid of many found in various publications. Although differing in words and some detail, most of them are quite similar overall. The following descriptions are intended to capture the spirit of the system.

---

[*] There is also a rating system used by Western guides using a scale from 1-10 that is usually reserved for large rivers having very heavy water, like the Colorado.

CLASS 1 **(Easy)**	The current is usually smooth and can easily be neutralized by correct backpaddling. Waves are up to six inches high and irregularities in the flow patterns are small. Passages are clear and the best course is obvious if obstacles are present. Ledges and abrupt drops are less then six inches. White water is not always visible.
Examples:	(1) Most of the North Branch of the Westfield from West Cummington to Cummington in low water. (2) Most of the White River near Rochester in low water.
CLASS 2 **(Medium)**	The rapids are moderately spaced, well defined, and may have numerous rocks necessitating some maneuvering. Waves are up to two feet high and are usually regular. The current can still be neutralized by backpaddling and the crosscurrents are not strong. There are abrupt drops in straightforward chutes and over easy ledges up to 2 feet. The boater must think a little about the best route, which is easily discovered. White water is usually visible.
Examples:	(1) Green River from West Leyden to the covered bridge in medium or low water. (2) Cold River from South Acworth to Vilas Pool in low or medium water.
CLASS 3 **(Moderately** **Difficult)**	The rapids are well defined, occurring frequently and often blending together. Waves can be irregular and choppy, up to 2.5 to 3 feet high, and crosscurrents can be strong enough to disorient a boat. Passages can be narrow, and obstacles can necessitate frequent maneuvering. The best path is not necessarily obvious at first, but is easily found upon inspection from land. Eddies are usually regular and hydraulics are strong enough to slow a moving boat noticeably. Abrupt drops are up to 2.5 to 3 feet. The current can be neutralized only with much effort. Moderate boat control is necessary and the boater must exert effort, both physical and mental. Open boats may require frequent stops for emptying water—a spray cover is useful.
Examples:	(1) Chickley River in high water. (2) Upper Ammonoosuc in medium water. (3) North Branch of the Westfield below Knightville Dam in medium water.

**CLASS 4
(Difficult)**

Class 4 rapids are not generally negotiable in an open boat paddled tandem without a spray cover. Rapids are long and involved, requiring precise boat control. Channels can be very narrow and twisted. Waves are irregular and powerful, up to 4 feet in height. Hydraulics can be strong enough to hold a boat, and crosscurrents can overturn a boat. Abrupt drops over ledges are up to 4 feet and souse holes can be 3 to 4 feet deep, with stopper waves following. Eddies have strong current differentials. Many times, only one route is practical and difficulties cannot be avoided. Class 4 rapids should be scouted the first time. Instant, often irreversible, decisions are necessary. The paddler cannot see rapids fully from the boat, and the full current cannot be neutralized by backpaddling. Eddy turns become very important, as does a strong, aggressive attitude. Boaters are continuously working and planning ahead. An Eskimo roll is very desirable for closed boaters.

Examples:

(1) East Branch of the Sacandaga in high water.
(2) Smith or Mad in medium water.
(3) Hudson River Gorge in medium water.
(4) East Branch of the Pemigewasset in medium water.

**CLASS 5
(Very Difficult)**

Class 5 difficulty is characterized by very long, heavy, violent, turbulent water. Passages are very complex with difficulty followed immediately by difficulty. The best passage is definitely not obvious, even when scouted from land. Abrupt drops are 4 feet and higher with waves being an irregular 4 feet and over. Visibility is usually severely limited by river obstructions, the extreme gradient, or large waves. Eddies and resting opportunites are few; the current is *very* powerful, and there is the possibility of whirlpools. The water level is critical for running. The utmost in skill and performance is required from the boater. Open boats are definitely out, and closed boaters should have a good roll.

Examples:

(1) Boreas River in high water.
(2) The Gorge and Staircase rapids on the lower Swift in high water.

CLASS 6 (Extraordinarily Difficult)	The difficulties of Class 6 water are those described in Classes 4 and 5, but exaggerated to the limits of navigability. The chief characteristic is that even the *expert paddler's life* is being risked. A single mistake in judgement or execution could lead to serious injury.

No example given.

Now, having read these descriptions, how many potential white water boaters have turned slightly timid? It is necessary to point out that not every individual characterization has to be present for a river, or rapids, to have a particular rating. Class 4 rapids don't all have souse holes with stopper waves followed by 4-foot abrupt drops. The descriptions given by each classification merely indicate what might be present. A Class 4 boater should be able to handle all the difficulties described, but it doesn't necessarily mean he'll get them thrown at him in one shot. Also two rivers, or rapids, may be rated identically, yet be totally different in character. For example, there are rapids that involve a great deal of maneuvering — as on the Swift and Boreas Rivers — but the water is not necessarily heavy. On the other hand, there are rapids where not many solid objects are visible, yet the water formations (waves, hydraulics, and souse holes) could stop a train. Finally, combinations of these two types can be blended together to form what is usually encountered. So, when inquiring about river classifications, give a little thought to the *kinds* of rapids.

Even with the above guidelines a certain amount of subjective judgment is called for in deciding what class a river is. A very conscientious effort has been made here to rate the difficulty objectively and consistently and not to be swayed by personal prejudice. Rivers, or sections of rivers, are classified according to the difficulty of the major rapids that are characteristic of that section. If a stretch has mostly Class 3 rapids with several harder spots, it is called a 3-3+, or a 3-4. If a river has intermittent Class 4 rapids, then it is usually rated a Class 4 even though the majority of the water may be Class 2. Rivers that are slightly harder or easier than a classification, but not enough to deserve another rating, are designated + or − respectively. In the situation where two numbers are used to designate difficulty, e.g., 3-4, the last number also represents a higher water classification. As the water level increases, the difficulty rating approaches the last number.

DRIVING DISTANCE

These columns state the approximate one-way driving distance in miles to the general area of the put-in. Mileage is measured from one of two origins: a) the intersection of the Massachusetts Turnpike and Route 128

(Boston), and b) the intersection of the Massachusetts Turnpike and Route 91 (Springfield). Mileages were obtained by measuring the appropriate distances on road maps and using the map key to convert these distances to miles. The values obtained were always rounded up to the nearest five-mile increment. In calculating the mileages tabulated here, highways marked with route numbers were always used, and interstate and turnpike routes were chosen when possible.

When figuring driving times, consider the types of roads you will be traveling and your own individual driving style. If you have to travel from Springfield to the Mad River in New Hampshire, for instance, the trip across New Hampshire will be longer in time than the mileage indicates because there are no direct, limited-access highways, as there are if you start out near Boston. All trips from Boston to Vermont were calculated using a Route 2-to-Route 91 itinerary.

IN

This column lists the name of a geographic location associated with the start, or put-in spot, of the trip. In most cases it is a town or community which may be quite unrecognizable as such from the road, such as Green River, Vermont, which consists of several houses and a covered bridge. Most of the smaller settlements do, however, show up on topographic maps. Some entries are route numbers and the specific starting spot can be found in the detailed description in Part II. A good road map is of invaluable aid in finding the small, out-of-the-mainstream towns that defy detection, or prompt the last-minute question, "Where the hell is . . .?"

OUT

This column lists the name of a geographic location associated with the take-out spot.

TRIP DISTANCE

This column lists the trip lengths in river miles. These distances were obtained by measuring the lengths of the trips on a topographic map and using the scale factor to convert the measured distance to miles. The trip lengths are "down the middle of the river" distances and do not take into account the distance paddled in maneuvering. These figures represent the minimum distance a boater must travel on the trip. If a river is particularly rocky, it is conceivable that the actual distance traveled could be significantly increased from the given value. Values for the shorter trips probably have a higher percentage error than those for longer trips because of the method of measurement. Also, rivers with a number of closely spaced meanders are more difficult to measure than those without. All other factors being equal, trips of 2-4 miles are considered short, whereas 12-13 miles is a long voyage for a single day's outing.

AVERAGE GRADIENT

This column gives figures for the average gradient of the river in feet per mile over the total length of the trip. These values were determined by counting the number of contour intervals that cross the river on a topographic map in its path from start to finish, subtracting one, multiplying the result by the contour interval distance stated in the map key, and then dividing this product by the trip length. For example, on the 6.8 miles of the Upper Miller's trip from South Royalston to Athol there are 23 contour intervals crossing the river and there is one interval for every 10 foot change in elevation for these particular topo maps. The river then drops a total of 220 feet in this distance, so the average gradient is 32 feet per mile.

Interpreting values of average gradient can be somewhat tricky. In general, the faster a river drops, the harder will be its rapids. However, there are enough exceptions to this rule to almost invalidate it. Not only must one consider how fast a river drops, but also, more important, *how* it drops. If the decrease in elevation is gradual, it is quite possible that there will be only a fast current with no meaningful rapids, e.g., trip A on the North Branch of the Westfield. If the riverbed drops abruptly in discrete spots, waterfalls result, or, what is more commonly encountered, ledges. Streams of the same average gradient can be completely dissimilar in character, depending on discharge rate, number of rocks, riverbed characteristics, etc. The Snake River that runs through Hell's Canyon in Idaho forms, in one place, standing waves measuring 15 feet from trough to crest, yet the average gradient is only about 9 feet per mile for this section ending at Lewiston. Even with this relatively low drop, the Snake (a *very* large river) is constantly boiling and churning as if hydraulic engineers were busily heating selected columns of water. To give an example somewhat closer to home, the Housatonic between the mouth of the Ten Mile and Gaylordsville has an average gradient of about 16 feet per mile, yet there are two very heavy rapids in this section. The Snake and Housatonic are excellent examples that illustrate how a low average gradient can be deceiving. This type of deception, however, usually occurs only on large rivers.

Most rivers covered in this guide have a gradient that averages over 30 feet per mile. This corresponds to a gradient of 0.57 percent or greater, whereas a drop of 100 feet per mile corresponds to a gradient of slightly less than 2 percent. To get a feeling for the magnitude of these numbers, consider Route 16 in the White Mountains as it comes south, down from Pinkham Notch. This road has a 9 percent gradient, as does the road that descends westward from the top of the Kancamagus Highway.

MAXIMUM GRADIENT

This column gives values for the maximum gradient in feet per mile encountered in a particular trip over a distance of ¼ - ½ mile or greater. These numbers were obtained by locating the section on a topographic map where the contour intervals crossing the river were judged to be the most closely spaced and then measuring the average gradient over this region in the manner already described. It is by no means the absolute maximum gradient to be encountered, since individual rapids will produce much larger drops. These larger gradients, however, are usually short-lived. In most cases, the maximum gradient values were not obtained for sections containing waterfalls or unrunnable rapids.

The figure for maximum gradient, when used with the corresponding value for average gradient, can give a much better picture of the trip than either does alone. For instance, the trip from West Cummington to Cummington on the North Branch of the Westfield has a maximum gradient that is not much higher than the average gradient, indicating a fairly uniform descent, which is the actual case. In the case of the Quaboag, the maximum gradient is almost three times that of the average, and one should expect a little excitement on some sections. Again, *how* the drop occurs is extremely important and no figures will tell the whole story as well as a personal visit.

SHUTTLE

This column gives the approximate one-way driving distance by a reasonable route from put-in to take-out. The routes are designed so that hard-surface roads can be used to a maximum. Those shuttle routes not obvious are easily found with the aid of a good road map. In several cases shuttle routes are detailed in Part II.

SCENERY

Scenery is rated according to the amount of civilization visible from the river. Rivers surrounded by an untouched wilderness are rated excellent, and ones with succeeding amounts of civilization visible are rated good, fair, or poor. Fair or poor ratings usually represent an undesirable stream in respect to scenery and can indicate pollution in an otherwise scenic river valley, e.g., the Upper and Lower Millers. The Hudson, Gale, Rapid, Green, White, Dead, and Boreas are judged to have excellent scenery.

RIVER LEVEL

In this column an attempt is made to indicate how various water levels relate to a river's canoeability. Using the existing Geological Survey gages present on most New England rivers and a collection of hand-painted ones, a correlation has been made between a specific gage reading and the corresponding level of the respective river as it pertains to canoeing. Gage readings are grouped according to five different classifications: TOO LOW, LOW, MEDIUM, HIGH, and TOO HIGH.

A gage reading in the TOO LOW column corresponds to a level at which a significant number of rapids on the trip do not have enough water in them for comfortable canoeing. This means that if boated, there would be a substantial amount of bottom-scraping, canoe-tugging, and cursing. It is a level which will do damage to the underside of a fiberglass boat or leave the rocks with a bright new aluminum-colored decoration. At this level, the main skill required is to figure out where most of the water is and to try to follow it. Proper paddle strokes are difficult to execute, and the boat is many times propelled by pushing off the bottom or off the rocks. In rocky streams, this level is to be avoided. Since most paddlers don't care for this sort of exercise, TOO LOW has been rated rather conservatively. Some trips are passable at this TOO LOW level, but they are not enjoyable.

The TOO HIGH rating indicates a level that should not be run by the *average* boater; even the experts will have difficulty, and it is questionable whether the risks are worth the thrills. Make even a small mistake and the consequences could be serious. If you consistently paddle this level, you are either used to it and are very good, or your wrappings are coming loose. Only a few rivers have a TOO HIGH rating. For many, this level is either inappropriate or it is not commonly found. Class 2, and sometimes Class 3 rivers, are occasionally washed out at this level. Rivers in flood stages are TOO HIGH.

The preferred levels are marked LOW, MEDIUM, or HIGH. LOW denotes a level that is pleasant but not challenging for a competent boater. On a Class 2 or 2+ river, LOW is a good level for beginner's instruction. The current is usually moderate and the difficulties can be handled by a moderate amount of thought and action. Wider sections of river may be shallow at this level. MEDIUM levels are those where an experienced boater will find a very smooth and challenging run. The dangers involved are not great, but they exist. Currents can be pushy and disorienting. Playing and practice are frequently done at this level. Intermediate and advanced instruction can also be carried out at this level.

HIGH levels are the precursors of TOO HIGH. The banks are not well defined at this level; trees and brush are knee-deep in water, making eddying-out difficult. Many eddies are washed out and those few remaining are difficult to negotiate. The current is frequently cluttered with floating debris such as trees, logs, bottles, ice (in season only), and broken boats. In general, HIGH is a level only for those who know what they are doing. A mistake isn't necessarily disastrous, although good boat control is needed. Rescues at this level are *very* tough, and the normal difficulty rating is usually increased. A HIGH level on a Class 4 river can be dangerous, while on a Class 2 river, it can just be unenjoyable for an experienced boater. At HIGH levels, the river is definitely the master, and the boater is just trying to stay even.

River level ratings depend on what type of boat is being used. What's MEDIUM for a kayak may easily be TOO HIGH for an open boat. What's MEDIUM for an open boat could be dull for a C-2 team. In the ratings, then, an effort has been made to rate the level for different types of boats. *A gage reading followed by a C indicates a level for a closed boat. A gage reading followed by an S indicates a level for an open boat, paddled solo. A gage reading with no other designation is for an open boat paddled tandem.*

Although there still are, and hopefully there will continue to be, a great number of open boaters, most white water enthusiasts today use closed boats of one type or another. The inherent disadvantage of open boats is that they can swallow huge amounts of water in hard rapids, which makes subsequent handling more difficult. Also, open boats cannot sustain a roll in the event of a dump. For these reasons, the tough rivers considered in this guide are given level ratings that apply primarily to closed boats. It should also be mentioned that there is a trend for some of the better open boaters to tackle Class 4 water. This is not, however, nor should it be, a practice for everyone. Class 4 open boating is difficult: it requires good judgment, skilled execution, and experience. Because it looks easy doesn't mean that it actually is.

The level ratings are designed for the intermediate to advanced paddler in a particular difficulty class. Although it is very difficult to define "intermediate" or "advanced" precisely, consider an intermediate to be someone who has had two years of experience paddling regularly with and learning from a competent white water group. An intermediate should be able to handle LOW levels, but will find HIGH levels to be very difficult. An intermediate open boater isn't necessarily an intermediate closed boater and vice versa.

To be absolutely correct, there exists a series, or range, of gage readings that would be appropriate for every level rating. In this book, *only the highest reading observed or reported for that hypothetical range is listed for the TOO LOW, LOW, MEDIUM, and HIGH levels; for the TOO HIGH level, the lowest value is given.* This convention is summarized in the following table.

TOO LOW	LOW	MEDIUM	HIGH	TOO HIGH
highest gage reading is given	*highest gage reading is given*	*highest gage reading is given*	*highest gage reading is given*	*lowest gage reading is given*

As a practical example of the above convention, consider the Contoocook River trip from Hillsboro to Henniker. The ratings are as follows:

TOO LOW	LOW	MEDIUM	HIGH	TOO HIGH
5.6	7.4C	9.0C	10.0C	10.0C

At any gage reading below 5.6, the river is too low for comfortable canoeing in any kind of boat. A level over 10 represents a run that's probably too difficult for all but the *best* boaters (or the nuts) since the river is in flood. At readings between 5.6 and 7.4, the level will be in the low range for closed boats; at readings between 7.4 and 9.0, the level will be in the medium range for closed boats. Gage readings between 9.0 and 10.0 represent a high level for the river. Not all rivers have a complete set of canoeability or LEVEL ratings. Where there is a void, the reader must use judgment.

As with difficulty, level assignments are somewhat arbitrary and subjective. The basis for classifications made here is the author's personal experience on these rivers or the advice of competent canoeists. An effort has been made to exclude personal prejudice and to represent things as they are. Not everyone will agree with these ratings, but they can at least serve as guidelines. Considerable care has been taken to make the ratings consistent from one river to another. If you find the LEVEL ratings to be too high or too low on several rivers, then take that into account when using the ratings on another river.

FLOOD

This column gives the minimum gage reading at which the river, or section of river, is officially considered to be in a state of flood. These values were obtained from the River Forecast Center in Hartford, Connecticut, which is a part of the National Weather Service. Flood stages are determined by on-site surveys in conjunction with local officials. Flood stage is that stage at which the overflow of the natural banks of the river begins to cause damage to any portion of the reach for which the gage is used as an index. Not all gages have been calibrated for flood. This level is to be avoided, since a boater will encounter difficulties not normally seen or previously experienced. *Canoeing rivers in flood is a matter of survival, not sport.*

The readings for the Sacandaga River were obtained from the Geological Survey's office in Albany and are values at which they consider the river to be very high, although not necessarily at a level the Weather Service would consider flood.

GAGE LOCATION

This column gives the locations for the various gages. Locations are usually listed as the name of the nearest town. Detailed descriptions of gage locations are found in Part II at the end of every trip description.

GAGE READINGS

The New England area is quite fortunate in having U.S. Geological Survey gaging stations on so many canoeable streams and rivers. Nearly every stream of medium size has a gage on it, and only the smaller, more obscure streams lack any such measuring devices. Although these gages were not placed for the purpose of helping itinerant canoeists, they can certainly be of help in determining the level of a river if the trouble is taken to use them. In Massachusetts, Vermont, and New Hampshire alone, there are over 200 of these gages. Sometimes jointly owned and maintained with the Corps of Engineers or the Weather Bureau, these gages are often located where only a minimum effort is needed to find and read them. Recognizable as tall, narrow, concrete structures by the riverbed itself, the gaging stations also have external staffs that are calibrated in tenths and sometimes hundredths of feet. These external staffs are quite rigid and extend into the water regardless of the level so as to allow a reading of the river's height to be taken. A completely automatic and separate measuring device inside the little blockhouse permits permanent record-keeping. This permanent record is either punched out onto paper tape at regular intervals (digital recorder) or plotted continuously on a graph (analog recorder). In some instances, the external staffs are not continuous, but instead are made up of several separate staffs which are still situated in the general area. As the water level changes, one section may become completely submerged; it then becomes necessary to find the next highest section.

A closeup view of the outside staff measuring high levels on the Contoocook at Henniker — Ray Gabler

The gage house and associated outside staffs on the Ashuelot at Gilsum. The diagonal staff is for high levels and the short vertical staff is for very low levels — Ray Gabler

Two types of external staffs are in general use today. The newest model has an enameled section with a white background upon which the calibrations are printed in black. Older gages, constructed from staffs of wood, have the calibrations in tenths marked by large metal staples. Both types can be mounted either vertically or on a slant. Some external staffs are attached directly to the gaging station itself; many can also be found a short distance away attached to concrete walls, trees, or bridges, etc. During snow and ice cover, some external gages are quite difficult to locate even if their general area is known.

In general, it is necessary to visit the gage site in order to obtain a gage reading. However, there are a selected number of gaging stations located throughout the area which are automated so that they can be monitored by the district office of the Geological Survey or by the U.S. Corps of Engineers. The name of this monitoring system is Telemark. This system was constructed to enable various government agencies to maintain constant and immediate contact with river discharge conditions, as is necessary in cases of flood, hurricane, or spring runoff. The network of gaging stations provides data that, when interpreted, allow the agencies to determine proper usage of existing flood-control facilities. The system can also be used for an additional purpose, to help boaters locate the best water. Some of the gaging stations in the system that are of interest to boaters are given below.

River	*Gage*
Ammonoosuc	Bath NH
Ashuelot	Hinsdale NH
Deerfield	Charlemont MA
Farmington	Tariffville CT
Farmington	Rainbow MA
Hudson	N. Creek NY
Mascoma	W. Canaan NH
Pemigewasset	Plymouth, Woodstock NH
Piscataquog	Goffstown NH
Sacandaga	Hope NY
West	Newfane VT
Westfield	Westfield MA
W. Branch Westfield	Huntington MA
W. Branch Farmington	New Boston MA
White	W. Harford VT

In addition to these gages, the Corps of Engineers has control over several dams that regulate the flow on a number of white water rivers. Dams that are of interest to boaters include:

Dam	*River*
Ball Mt	West (NH)
Birch Hill	Millers (MA)
Blackwater	Blackwater (MA)
Knightville	Westfield (MA)
Littleville	Westfield (MA)
Otter Brook	Otter Brook (NH)
Surry Mt	Ashuelot (NH)
Tully	Millers (MA)

With these sources of information, it is possible for boaters to know a lot about water levels before stepping out of the house. This makes matters really convenient because you can decide whether or not to go boating, and where, without having to drive to an empty river. The Corps of Engineers in Waltham,

Massachusetts has agreed to give river information to boaters. The Boston Chapter of the Appalachian Mountain Club (AMC) is responsible for coordinating all activities with the Corps of Engineers. Every Friday during boating season, someone from the Boston AMC calls the Corps for river gage readings. That person then communicates these readings to any interested party. This single call method of coordinating with the Corps reduces the number of phone calls they receive, and minimizes interruptions to their normal work. Any club that is interested in obtaining gage readings should have *one* member call the Boston AMC coordinator for this information. If you want a particular gage reading and you don't know if the Boston AMC coordinator asks for it from the Corps, call the coordinator before Friday to insure this request is made. To find out who the current Boston coordinator is, call the AMC headquarters on Joy Street in Boston.

For the Hudson gauge, call the special weather service at the Albany airport (518-869-7891). Wait until the recorded message stops, then ask the person who answers for the gage reading at North Creek. Although the Hope gage on the Sacandaga is not really automated, an observer relays daily readings to the Hudson River-Black River Regulating District office on State Street in Albany (518-465-3491).

RATING TABLES

One of the most important characteristics of a gaging station is the rating table which relates the actual flow discharge to the gage reading. Rating tables for the rivers with U.S.G.S. gages are given in the Appendix. These tables appear for two reasons. First, they allow one to determine the actual (or approximate) amount of water flowing in a particular section of river if the gage reading is known. It is the amount of water flowing in a river that is the important factor here, not the arbitrary scale reading of the gage. It is best to think in terms of cubic feet per second (CFS) if possible, as this is the most widely used scale for measuring flow. Since gages are on all sorts of rivers with different riverbed characteristics, the same readings on different gages are completely independent of one another. Each river needs its own rating table, and, as riverbeds change, so must the rating tables if they are to be current. The tables included in the Appendix, obtained from the U.S.G.S., are those in current (1980) use. When they are no longer current, the tables in the Appendix will still serve as a good approximation.

The second reason for presenting the rating tables is to allow the boater to compare a familiar river with a less familiar one. Knowing the average sizes or cross-sectional areas of two rivers, it is possible to determine a necessary gage reading on the unfamiliar river for a particular type of boating. An example should make this clear. Let us assume that the West Branch of the Westfield and the Saxtons River are roughly the same size and have the same gradient. Let us also assume that 800 CFS makes the Westfield an enjoyable run. A similar amount of water in the Saxtons should then produce a similar type of trip. By using the Saxtons River rating table to match discharges, one finds that a gage reading of 5.7 is needed on the Saxtons.

As a guide for relating gage readings to the amount of water in a river, Graph 1 illustrates reading vs. CFS discharge rate. This particular graph is the rating table for the gage on the Saxtons River, but it is representative of most curves. Note that the discharge rate is roughly proportional to the square or cube of the gage reading. This is due to two closely related facts. First, the velocity of water in a river increases rapidly as the gage reading, or depth, increases. Second, it takes more water to add one inch of depth to a river when that river is at a high level compared to when it is at a low level because the banks slope outward and do not go straight up. The reader should realize that at higher levels, a small increase of 0.1 or 0.2 feet in gage height means considerably more water than the same increase at low river levels. To illustrate, let us consider the Saxtons River as a specific example. The difference in discharge rates between levels of 3.0 feet and 3.2 feet is 17 CFS, whereas it is 120 CFS between levels of 6.0 feet and 6.2 feet. Even though the change in gage levels is 0.2 feet in both cases, there is much more water involved when the river is high compared to when it is low.

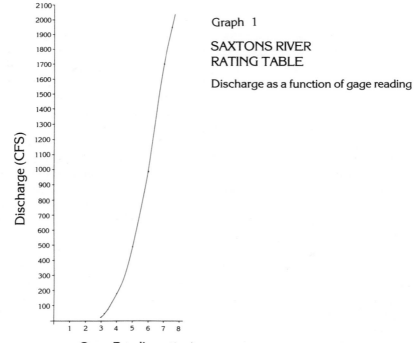

Graph 1

SAXTONS RIVER
RATING TABLE

Discharge as a function of gage reading

Discharge (CFS)

Gage Reading (feet)

VARIATION OF FLOW WITH SEASON

Graphs 2, 3, and 4 show the variation in discharge rate with season. These graphs represent the flow discharge at the designated sites averaged over 39 years for every day of the year. This type of averaging smoothes out anomalous flows due to abnormal weather conditions like floods and droughts. If you made a similar graph but included just one year's data, there would be many more peaks and valleys due to the effects of regional rains, etc. When one looks at these graphs, several points stand out. First, as expected, the largest flow is during the spring runoff, and the peak discharges of the three rivers are displaced from one another because of their locations. The south-ernmost rivers have high water prior to the northern river, which makes sense because the north stays colder longer, delaying the spring thaw. The next period of highest flow occurs in the late fall when the discharges are maybe one-fourth to one-third that of the spring peaks. An extensive canoeing pro-gram could then continue well into November and possibly into December. Boating may not be terribly exciting then, but at least it's possible. Also, the water will be warmer in the fall than in the spring because rain rather than cold snow runoff brings the rivers up in the fall.

Graph 2

WEST BRANCH WESTFIELD

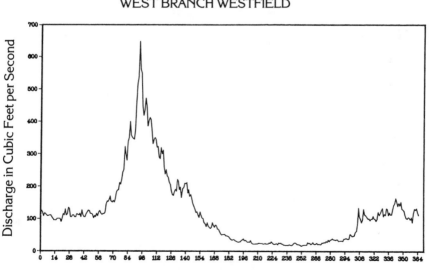

Day of Year

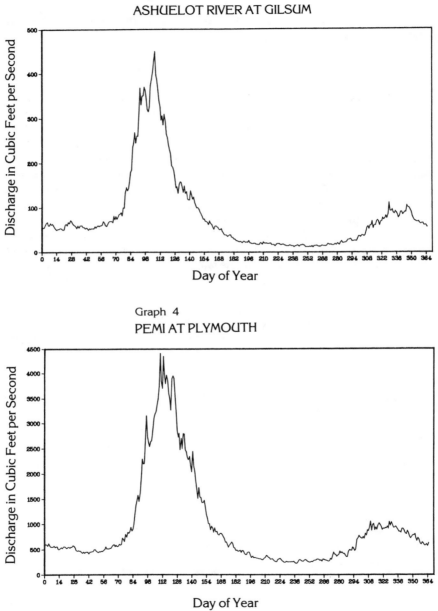

Graph 3
ASHUELOT RIVER AT GILSUM

Graph 4
PEMI AT PLYMOUTH

One last point to make is that it is possible to create a canoeing schedule where the probability of catching a river at its peak flow can be maximized. This is done by using graphs similar to the ones shown. Each river has a two- or three-week period where the probability (based on historical data) of finding good water is high, and that's when those canoeing trips should be scheduled.

Safety

Although it appears nearly last in this discussion, safety should be practiced first on river trips. There is little doubt that white water canoeing *can be dangerous.* The only reason more individuals don't sustain serious injury is due to the rigorous safety standards of organized clubs. Most tragic accidents on white water usually involve the violation of a major safety law(s)—canoeing alone, without a lifejacket, in flood stage, above one's ability, in cold water, etc.

Through many years of experience, boaters of the American Whitewater Affiliation have developed a formal safety code which covers the points of safety to be observed. It was first adopted in 1959, revised in 1963, and again in 1974. The following is the AWA Safety Code:

I. PERSONAL PREPAREDNESS AND RESPONSIBILITY

1. BE A COMPETENT SWIMMER with ability to handle yourself underwater.

2. WEAR A LIFEJACKET.

3. KEEP YOUR CRAFT UNDER CONTROL. Control must be good enough at all times to stop or reach shore before you reach any danger. Do not enter a rapids unless you are reasonably sure you can safely navigate it, or swim the entire rapids in the event of a capsize.

4. BE AWARE OF RIVER HAZARDS AND AVOID THEM. Following are the most frequent KILLERS:

   a. *HIGH WATER.* The river's power and danger and the difficulty of rescue increase tremendously as the flow rate increases. It is often misleading to judge river level at the put-in. Look at a narrow critical passage. Could a *sudden* rise from sun on a snow pack, rain, or a dam release occur on your trip?

   b. *COLD.* Cold quickly robs one's strength, along with his will and ability to save himself. Dress to protect yourself from cold water and weather extremes. When the water temperature is less than 50° F., a diver's wetsuit is essential for safety in the event of an upset. Next best is wool clothing under a windproof outer garment such as a splash-proof nylon shell; in this case, one should also carry matches and a complete change of clothes in a waterproof package. If, after prolonged exposure, a person experiences uncontrollable shaking, or has difficulty talking and moving, he must be warmed immediately by whatever means available.

c. *STRAINERS.* Brush, fallen trees, bridge pilings, or anything else which allows the river currents to sweep through but pins boat and boater against the obstacle. The water pressure on anything trapped this way is overwhelming, and there may be little or no white water to warn of danger.

d. *WEIRS, REVERSALS, AND SOUSE HOLES.* The water drops over an obstacle, then curls back on itself in a stationary wave, as is often seen at weirs and dams. The surface water is actually going UPSTREAM, and this action will trap any floating object between the drop and the wave. Once trapped, a swimmer's only hope is to dive below the surface where the current is flowing downstream, or to try to swim out the end of the wave.

5. BOATING ALONE is not recommended. The preferred minimum is three craft.

6. HAVE A FRANK KNOWLEDGE OF YOUR BOATING ABILITY. Don't attempt waters beyond this ability. Learn paddling skills and teamwork, if in a multiple-manned craft, to match the river you plan to boat.

7. BE IN GOOD PHYSICAL CONDITION consistent with the difficulties that may be expected.

8. BE PRACTICED IN ESCAPE from an overturned craft, in self rescue, in rescue and in ARTIFICIAL RESPIRATION. Know first aid.

9. THE ESKIMO ROLL should be mastered by kayakers and canoers planning to run large rivers and/or rivers with continuous rapids where a swimmer would have trouble reaching shore.

10. WEAR A CRASH HELMET where an upset is likely. This is essential in a kayak or covered canoe.

11. BE SUITABLY EQUIPPED. Wear shoes that will protect your feet during a bad swim, or a walk for help, yet will not interfere with swimming (tennis shoes recommended). Carry a knife and waterproof matches. If you need eyeglasses, tie them on and carry a spare pair. Do not wear bulky clothing that will interfere with your swimming ability when waterlogged.

II. BOAT AND EQUIPMENT PREPAREDNESS

1. TEST NEW AND UNFAMILIAR EQUIPMENT before relying on it for difficult runs.

2. BE SURE CRAFT IS IN GOOD REPAIR before starting on a trip. Eliminate sharp projections that could cause injury during a swim.

3. Inflatable craft should have MULTIPLE AIR CHAMBERS, and should be test-inflated before starting trip.

4. HAVE STRONG, ADEQUATELY SIZED PADDLE OR OARS for controlling the craft, and carry sufficient spares for the length of the trip.

5. INSTALL FLOTATION DEVICES in non-inflatable craft, securely fixed and designed to displace as much water from the craft as possible.

6. BE CERTAIN THERE IS ABSOLUTELY NOTHING TO CAUSE ENTANGLEMENT when coming free from an upset craft, i.e., a spray skirt that won't release or tangles around legs; lifejacket buckles or clothing that might snag; canoe seats that lock on shoe heels; foot braces that fail or allow feet to jam under them; flexible decks that collapse on boater's legs when a kayak is trapped by water pressure; baggage that dangles in an upset; loose rope in the craft or badly secured bow/stern lines.

7. PROVIDE ROPES TO ALLOW YOU TO HOLD ONTO YOUR CRAFT in case of upset, and so that it may be rescued. Following are the recommended methods:

   a. KAYAKS AND COVERED CANOES should have 6-inch diameter grab loops of 1/4-inch rope attached to bow and stern. A strern painter 7 or 8 feet long is optional and may be used if properly secured to prevent entanglement.

   b. OPEN CANOES should have bow and stern lines (painters) securely attached, consisting of 8 to 10 feet of 1/4- or 3/8-inch rope. These lines must be secured in such a way that they will not come loose accidentally and entangle the boaters during a swim, yet they must be ready for immediate use during an emergency. Attached balls, floats, and knots are *NOT* recommended.

   c. RAFTS AND DORIES should have taut perimeter grab lines threaded through the loops usually provided.

8. RESPECT RULES FOR RAFT CAPACITY and know how these capacities should be reduced for white water use. Liferaft ratings must generally be halved.

9. CARRY APPROPRIATE REPAIR MATERIALS; tape (heating duct tape) for short trips and complete repair kit for wilderness trips.

10. CARTOP RACKS MUST BE STRONG and positively attached to the vehicle, and each boat must be tied to each rack. In addition, each end of each boat should be tied to the car bumper. Suction cup racks are poor. The entire arrangement should be able to withstand all but the most violent vehicle accident.

III. LEADER'S PREPAREDNESS AND RESPONSIBILITY

1. **RIVER CONDITIONS.** Have a reasonable knowledge of the difficult parts of the run or, if an exploratory trip, examine maps to estimate the feasibility of the run. Be aware of possible rapid changes in river levels and how these changes can affect the difficulty of the run. If important, determine the approximate flow rate or level. If the trip involves important tidal currents, secure tide information.

2. **PARTICIPANTS.** Inform participants of expected river conditions and determine if the prospective boaters are qualified for the trip. All decisions should be founded on group safety and comfort. Difficult decisions on the participation of marginal boaters must be based on total group strength.

3. **EQUIPMENT.** Plan so that all necessary group equipment is present on the trip; 50- to 100-foot throwing rope, first aid kit with fresh and adequate supplies, extra paddles, repair materials, and survival equipment if appropriate. Check equipment as necessary at the put-in, especially: lifejackets, boat flotation, and any item that could prevent complete escape from the boat in case of an upset.

4. **ORGANIZATION.** Remind each member of individual responsibility in keeping group compact and intact between the leader and sweep (capable rear boat). If group is too large, divide into smaller groups, each of appropriate boating strength, and designate group leaders and sweeps.

5. **FLOAT PLAN.** If the trip is into a wilderness area, or for an extended period, your plans should be filed with appropriate authorities or left with someone who will contact them after a certain time. Establishment of checkpoints along the way, at which civilization could be contacted if necessary, should be considered. Knowing the location of possible help could speed rescue in any case.

IV. IN CASE OF UPSET

1. EVACUATE YOUR BOAT IMMEDIATELY if there is imminent danger of being trapped against logs, brush, or any other form of strainer.

2. RECOVER WITH AN ESKIMO ROLL IF POSSIBLE.

3. IF YOU SWIM, HOLD ON TO YOUR CRAFT — it has much flotation and

is easy for rescuers to spot. Get to the upstream end so craft cannot crush you against obstacles.

4. RELEASE YOUR CRAFT IF THIS IMPROVES YOUR SAFETY. If rescue is not imminent and water is numbing cold, or if worse rapids follow, then strike out for the nearest shore.

5. EXTEND YOUR FEET DOWNSTREAM when swimming rapids to fend against rocks. LOOK AHEAD. Avoid possible entrapment situations; rock wedges, fissures, strainers, brush, logs, weirs, reversals, and souse holes. Watch for eddies and slack water so that you can be ready to use these when you approach. Use every opportunity to work your way toward shore.

6. If others spill, GO AFTER THE BOATERS. Rescue boats and equipment only if this can be done safely.

Cold Water — Drowning Hypothermia*

The early spring (and late fall) boater is confronted with the most serious hazard a river has to offer — COLD WATER. Sudden immersion in cold water may produce serious consequences. Immediate death may result from drowning or cardiac arrest. Exposure to cold water for anywhere from a few minutes to an hour or more leads to hypothermia, a lowering of the core body temperature (measured by rectal, not oral, thermometers). Unconsciousness and, consequently, a risk of drowning occur whenever the core body temperature is reduced appreciably. A further lowering of body temperature inevitably leads to death.

Sudden immersion in water colder than 50° F causes immediate and intense difficulty in breathing. Gasping and inability to control breathing may cause the dunker to panic within a few minutes following a cold water tipover. Whenever the head of a gasping dunker is covered by a wave (even if only for a few seconds), he is likely to inhale water and possibly drown. Gasping and rapid breathing can produce hyperventilation within ten seconds, and this can lead to unconsciousness. Consequently the likelihood of drowning increases because even the best lifejacket cannot be expected to hold the head of an unconscious dunker above the water in turbulence.

In addition to the breathing difficulties associated with cold water immersion, the dunker's body rapidly loses heat at a rate MUCH FASTER THAN THE DUNKER REALIZES. Depending on skin thickness and the

* This section is reprinted from *Whitewater: Quietwater, A Guide to the Wild Rivers of Wisconsin, Upper Michigan and NE Minnesota* by Bob and Jody Palzer, Fourth Edition, copyright 1980. Reproduced with permission of the publisher, Evergreen Paddleways, 1416 21st St., Two Rivers, WI 54241.

amount and kind of protective clothing worn, the dunker's core body temperature drops from a normal value of 98.6° F to about 96° F within two to ten minutes. At these low core body temperatures, useful work becomes difficult and often impossible. Consequently the dunker is unable to swim or in any other way assist in his own rescue. A drop in core temperature to 88° F leads to unconsciousness, and a further drop to about 77° F usually results in immediate death.

To understand hypothermia, consider how the body gains, conserves, and loses body heat. The human body is heated in two ways, externally and internally. Heat is produced internally by the oxidation of food (metabolism) which occurs at one rate in a resting individual and at a higher rate during exercise. Shivering is an involuntary form of muscular exercise that produces heat at the expense of body food stores. The body can also receive external heat from the surroundings (sun, fire, another body, etc.). Body heat is conserved by the insulative layer of subcutaneous fat and by constriction of the blood vessels to the extremities. Body heat is lost by exhalation of warm air, by sweating, and by radiation, convection, and conduction from exposed body surfaces. To the boater, convection and conduction are the most important means of heat loss.

Rapids boaters are generally subjected to continual wetting even in the absence of tipovers. Wet clothing drains heat from a boater's body at an alarming rate because water conducts heat away from the body at a rate twenty times faster than air. Complete immersion is an even more serious hazard to the boater. Within seconds after immersion, clothing loses its insulative properties; these are due primarily to trapped air which prevents heat loss by reducing convection. Unless a garment is waterproof, wet clothing cannot prevent the rapid circulation of cold water past the skin (convection). Within minutes, the skin temperature of the dunker who is not wearing waterproof clothing will drop to within a few degrees of the water temperature. The dunker's body automatically responds by cutting off blood circulation to the skin in an attempt to conserve heat. Reduced blood circulation prevents the transport of energy stores required by the muscles to perform voluntary work. Thus voluntary movement of the extremities becomes increasingly difficult, sapping body strength and incapacitating the dunker.

Muscular activity on the part of the dunker creates a dilemma. Muscular activity is needed for the dunker to rescue himself, but it also increases the rate of heat loss since blood forced to the extremities is cooled rapidly by the cold water. In short, the dunker should ALWAYS GET OUT OF COLD WATER AS QUICKLY AND WITH AS LITTLE EXERTION AS POSSIBLE. If feasible, he should lie on his back in a feet-first position and let the current carry him downstream as he works his way to the shore. He should forget about saving his boat and rescuing other equipment. Unconsciousness can occur within five minutes after immersion in cold water. Scientific experiments and personal experience have shown that even good swimmers cannot swim a distance of 200 yards when the water temperature is less than 40° to 45° F. Differences in skin thickness, basal metabolic rate, and other physiological parameters produce variability in the timing of an individual's response to immersion in cold water. Therefore, the following table should be considered only as a rough guide. We can't tell you how long YOU will last in cold water. Physical fitness is not an indicator. The physically fit person may be the first to go because body fat provides protection against cold.

Water Temperature	Useful.Work	Unconscious
32.5°	less than 5 minutes	less than 15 minutes
40°	7.5 minutes	30 minutes
50°	15 minutes	60 minutes
60°	30 minutes	2 hours
70°	45 minutes	3 hours

Adapted from Davidson, A.F. "Survival The Will To Live," AWA Vol. XII, No. 1, Summer, 1966.

Protection against the serious consequences of cold water immersion may be obtained either by acquiring a substantial layer of body fat or by wearing protective clothing. The above table lists estimates for a person clothed appropriately for the air temperature. Wearing waterproof outer layers extends these times somewhat, but wearing a neoprene wetsuit extends them considerably. Woolen underwear worn under a wetsuit (and lifejacket) provides the best protection against cold water that we know of.

Paddlers who have been wearing wet clothing for several hours may experience a mild case of hypothermia. This causes a reduction in the times listed in the above table. Consequently, be on the alert for signs of hypothermia in yourself and your companions. Symptoms include a lack of coordination, thickness in speech, irrationality, blueness of skin, dilation of pupils, and decrease in heart and respiratory rate. These symptoms are similar to those of a drunken person, and the judgment of a hypothermic boater is generally equally irrational.

Treatment: At least two persons in every boating party should be versed in mouth-to-mouth resuscitation and external cardiac massage, so that if one of the first-aiders becomes the victim there will be another available to provide assistance. Whenever a boater has been in the water for more than ten minutes, hypothermia is almost certain. Under these conditions artificial respiration should be given only if breathing has stopped completely, and then it should be applied "at no more than half the normal rate."[*] Otherwise the would-be rescuer may cause the victim to become hyperventilated since during hypothermia metabolism is slower and less oxygen is consumed.

[*] Keatinge, W.R. "Survival in Cold Water: The Physiology and Treatment of Immersion Hypothermia and of Drowning". Blackwell Scientific Publications, Oxford, 1969, p. 73.

The hypothermic victim must be supplied with external heat as rapidly as possible. Removal of wet clothing facilitates heating. If the victim has experienced only a relatively mild case of hypothermia, replacement of wet clothing with dry apparel usually suffices. However, in instances of extreme hypothermia, external heat must be applied since the victim is incapable of generating enough heat himself regardless of the amount of dry clothing supplied. A bath of warm water is the best source of external heat, but since it is not generally available other heat sources should be used. A fire or another warm body can be used if necessary.

SELF RESCUE

Self rescue is a technique every boater should practice. In rapids, especially difficult rapids, a swimmer cannot depend on others to get him out: it may take all they can manage to keep themselves upright. Many times complete rescue has to wait for calmer waters. If you dump and leave your boat, go immediately to its upstream side, then decide how to work your way to the nearest safe spot. If you happen to see another person following a boat down some rapids, you should obviously try to help him. If you can't do anything constructive for one reason or another, at least stay close and talk to the person. This will not make him feel any drier, but it will help morale, and that's important. As a swimmer, you may draw so much conversation that you will feel obliged to serve tea. You will also get contradictory advice, so use your head to sort things out — that's why your helmet is protecting it. As a rescuer, remember that the cardinal rule is to *go after people first and equipment second.*

ORGANIZED vs. SOLO BOATING

There is much to be said for paddling a calm, serene lake by yourself, enjoying a closeness with nature; however, when it comes to white water, the situation changes. Because of white water's inherent dangers, a party of three boats is considered the minimum number for a trip. If you get into trouble, you'll be glad of the extra help. People do boat alone, but it is definitely not recommended, because when trouble strikes there is nobody around to help.

Although organized groups can have too many committee meetings, recipe exchanges, and forms, they do offer an excellent way to learn the sport safely. Beginners can get instruction under controlled conditions from people who have more experience, and there's always a chance to talk about techniques and mistakes. It is possible to learn from books, but only to a certain point. Canoeing is best learned by doing. Organized clubs may not be the most efficient or the fastest way to learn, but they do offer some advantages. If you are a beginner and want to learn, consider joining one of these clubs. A list of some canoeing clubs in New England is presented in the Appendix.

Lifejackets

Anyone who has suddenly and unexpectedly been thrown from a boat into the water knows what a shock it can be — complete disorientation, momentary paralysis due to an instantaneous change in temperature, and realization that you must now think actively about how and when to breathe. The first few seconds in the water are panic as you try to reorient and to fight your way back to the surface. Occasionally the direction of the surface is none too clear, and, since you haven't prepared for this dunking, your air supply seems to be used up at an unusually rapid rate, causing even deeper panic. This picture is true for all sorts of boaters, from those who chase the America's Cup to those who go out on a lake for a Sunday paddle to white water canoeists. White water canoeists, however, have a problem that is somewhat peculiar to their sport. Once you manage to resurface, the current thrashes you about like a toy: you may be bashed against rocks, tree branches, your own boat, floating debris, and perhaps even some ice. In addition, if the current has any force to it, you'll be intermittently submerged as standing waves wash over your head and souse holes rotate you around and around like a washing machine. Now, this is all aggravation you don't really need, and the point to make here is a rather obvious one. Since you have a whole host of problems to contend with while in the water, you don't want to exacerbate the situation by having to worry about your natural buoyancy or lack of it. This is the single best argument for wearing lifejackets when canoeing white water. Among all your other problems, you don't have to worry about buoyancy or your ability to resurface if circumstances permit.

Competent white water boaters have a simple rule about lifejackets, and that is, they don't paddle without them. No exceptions. Jackets should always be used, especially on Class 1 and Class 2 water, even though it doesn't look menacing. People who boat Class 3 and 4 water are usually well aware of potential dangers, but easier water can lull you into a false sense of security, that is, until you go in and get swept downstream. Nobody is immune to problems in the water; adults and children alike have the same basic obstacles to contend with, so *everybody* should be properly equipped. A lifejacket does not guarantee your safety, but it will help. If you have ever been underwater, and felt the tug of your jacket pulling you to the surface, you'll never again question its usefulness, even though it sometimes seems a nuisance and a bother.

There are two other aspects about lifejackets that are seldom emphasized, but are important. The first is their ability to keep you warm, and the second is their capacity to cushion collisions. If you have a vest-type jacket with flotation on both front and back, you have another layer of insulation between you and the environment, whether it be gaseous or liquid. The vest models surround the core part of your body, acting as a kind of sleeveless wetsuit top which gives added protection against hypothermia or plain old chills. Some jackets are comfortable enough to be worn around camp or even while sleeping on a particularly cold night. Depending on the fit and style of your jacket, it can add measurably to your warmth.

If you make the transition from boater to river swimmer, lifejackets also provide extra padding when you come into contact with rocks, ledges, trees, etc. These contacts are frequently forceful, made without reference to proper etiquette, and broken ribs and bruises result. With a jacket on, the impact doesn't hurt as much, and again, if there's padding on the back, you're protected all around while being dragged over a boulder garden.

In 1973, Congress passed the Boating Safety Act which gave the Coast Guard jurisdiction over recreational boaters while on all navigable inland waterways. One of the results is that the Coast Guard now publishes and enforces regulations concerning lifejackets. These include the construction of jacket materials, the jacket's performance, and the carrying of jackets while boaters are on the water. The Coast Guard's jurisdiction encompasses boaters of all types, not just white water canoeists. In fact, white water canoeists represent a rather small and specialized subgroup of boaters. The Coast Guard has classified lifejackets into five groups which are described in their publication, *Federal Requirements for Recreational Boaters (CG-290)*, and they now use the term personal flotation device (PFD) to describe not just lifejackets but other devices not designed to be worn by the user as well. The five classifications and their descriptions are given below.

Type I

A Type I PFD has the greatest required buoyancy. It is designed to turn most unconscious persons in the water from a face-down position to a vertical or slightly backward position. The adult size provides a minimum buoyancy of 22 pounds and the child size provides a minimum of 11 pounds. An example is the jackets worn by commercial rafters in the Grand Canyon.

Type II

A Type II device is designed to turn its wearer in a vertical or slightly backward position in the water. The turning action is not so pronounced as with a Type I, and a Type II will not turn as many persons under the same conditions as will a Type I. An adult-size Type II provides a minimum buoyancy of 15½ pounds. The medium child size provides a minimum of 11 pounds, and the infant's size provides a minimum buoyancy of 7 pounds. Examples are the jackets commonly known as horse collars.

Type III

A Type III PFD is designed to put the wearer in a vertical or slightly backward position. While a Type III has the same buoyancy as a Type II, it has a smaller turning moment. It can, however, allow for greater wearing comfort. An example is the vest-type jacket.

Type IV

A Type IV device is one designed to be grasped and held by the user until rescued, as well as thrown to a person in the water. It is not designed to be worn. An example is a ring buoy or buoyant cushion.

Type V

A Type V device is one designed for a specific and restricted use.

The Coast Guard publication continues: *"All PFD's that are presently acceptable on recreational boats fall into one of these designations. All PFD's shall be U.S. Coast Guard-approved, in good and serviceable condition, and of an appropriate size for the persons who intend to wear them.*

All recreational boats less than 16 feet in length and all canoes and kayaks of any length must have one Type I, II, III, or IV PFD aboard for each person. Type I, II, and III devices must be readily accessible to all persons on board. The Type IV device shall be immediately available for use. PFDs must be Coast Guard-approved."

In developing the above definitions and regulations, the Coast Guard has tried to cover all types of boating activities with a small number of simple rules that can be used across the United States. If they tried to regulate each type of boating activity individually, the result would be a legal nightmare: exemptions in one area might be used as a basis for a successful exemption in another area where in fact they do not apply but where the law as written nevertheless permits application. The Coast Guard wishes to protect the casual boater, not the expert, and there are many casual boaters. The Coast Guard therefore is the dominant influence on the manufacturing quality of PFDs. For instance, the Coast Guard must approve not only the finished product, but also the basic raw materials, to insure they will withstand rough use. As a rule, this is something the actual user cannot or will not do. A white water canoeist is usually very critical in choosing a lifejacket. Most people are not.

Boaters should not take these regulations lightly. First, they make overall sense. There are situations peculiar to individual sports, such as white water canoeing, where the definitions of the PFDs don't work or don't seem logical, but in general the right idea is there. Second, people in New England have been stopped and fined for not having Coast Guard-approved jackets even though they were wearing lifejackets. Admittedly this happens rarely, but it does happen.

Dams

Although dams aren't as plentiful as rocks on New England rivers, if you paddle here for any length of time you'll soon find yourself approaching one. Many have been built for the power or processing needs of local industry. You'll find dams on all sorts of rivers and in various stages of function and repair. There are basically two types of dams: each has its own peculiarities and dangers. The first type has water going under the dam; the second has water going over the top. The latter dams are more numerous and more dangerous.

Dams that can regulate the flow of water present relatively few problems. They are usually quite large and easily visible at a distance. In addition, the pools behind them are frequently marked somehow to indicate how close you can approach without risk of being sucked in and run through a turbine. If nothing else, the maintenance personnel may spot you and invite you to get off their artificial lake. On the downstream side, the effluent can come out with tremendous force, so it is best to put your boat in the water well downstream. Also, portaging one of these monsters can turn out to be a major project.

The type of dam where the water flows over the top in an unregulated fashion can present several subtle dangers. First these dams are not always easy to spot since the current can continue right up to the edge of the dam itself. If you don't see it in time, over you go, or at least you may have a few frantic moments paddling your way back upstream. There are three clues that may indicate an upcoming dam: slack or dead water where you don't really expect it, a regular line across the river ahead, and the abrupt disappearance of the river. Although it may seem obvious that you should be careful of going over a dam by accident, and that you should be observant, people do manage to fall over them anyhow.

Frequently dams are run on purpose and in this instance you can encounter a second danger—hydraulics. The water formation at the base of a dam is called a hydraulic. It is a circulating current that is actually going upstream at the surface. A hydraulic is what keeps logs rolling around and around at the base of some large dams; it can do the same thing to you and your boat. The height or steepness of a dam does not always indicate how strong a hydraulic will be. The shape of the dam's base and of the river bottom, along with the amount of water going over the dam, all contribute to make a strong or weak hydraulic. It is often impossible to tell how strong a hydraulic is just by looking at it, a fact which new boaters should keep in mind. Some relatively low dams have vicious hydraulics while taller ones have innocuous ones. There are no hard and fast rules for predicting how a hydraulic will behave. Each must be considered separately. *If you have any doubts at all, don't run the dam.*

SCOUTING

Scouting is examining rapids from shore, prior to running or portaging, to note the best route and potential hazards. Although it sounds rather easy, good scouting can actually be quite tricky, due to the fact that, while paddling the rapids, you have a different view than you did on shore. So, once you have determined a route from the sidelines, walk upstream and squat down to river level to see how things will look when you're in the boat. If the route is still obvious, you are ready to run. If the route is no longer clear, re-evaluate the whole situation. Also, be aware that rapids can frequently paddle harder than they look.

Scout *all* rapids or blind curves that you are unsure of, even if others in the group don't. Make sure you look at the whole rapids, not just the start. A channel that's clear at the beginning can occasionally end up going under an impassable ice shelf. Rivers with snow and ice cover on the banks deserve very special attention, even from experts. Stay away from people who boast, "I never scout anything." They can get both of you killed. Scouting rapids from your boat by moving from one eddy to another is a common pratice among experienced boaters, but one not recommended for beginners. Last, don't become complacent or forgetful about scouting as you gain skill: you are probably tackling more difficult water and the rapids may tend to become people-eaters.

WARMUP

As with any athletic activity, adequate warmup prior to boating is important. White water canoeing usually happens when the weather is less than warm, when muscles and joints tend to creak more. It is common sense to loosen up before getting into your boat and blasting down to the first rapids. Everybody will have a different routine and that's fine. Some go through an elaborate set of exercises that systematically limber up all movable parts. Others just stand around and shiver. At least move your arms a little if for no other reason than to keep the vultures at a safe distance.

Hand Signals

Communication between people on a river is difficult under most circumstances. If there's any appreciable current at all, the noise of the moving water drowns out all but the loudest yelling. Where there are rapids, it takes great effort to be heard unless your intended listener is close. If the boats in your trip are strung out, the whole situation worsens: the only effective communication is visual. In rescues, teamwork is essential, but it's impossible to work together without knowing what the people on the other side of the river are going to do. Also, once a correct route is determined, how does the leader communicate it to the rest of the group?

To overcome these sorts of problems, a group of competent paddlers from all over the United States formed a committee to devise a standard set of hand signals for use on the river. These signals don't cover all possible situations. They do cover those cases where it is extremely important that there be mutual understanding among a group. These signals are for the most common situations you will encounter. You'll have to invent your own adaptations for exceptional cases.

The committee used these criteria to devise the standard signal system:

1) Number of signals should be kept to a minimum;
2) Signals should convey only essential information;
3) Signals should be highly visible;
4) Their meaning should be unambiguous, even at a distance;
5) Signals can be executed both in a boat and standing on shore;
6) Signals can be executed facing both toward and away from a group; and,
7) Signals should be simple, easily remembered, and easily taught.

The four visual and one auditory signal in this system, their execution, and their meanings are given below:

Signal   **Execution**

Help/Emergency
Help is needed immediately.
Come as fast as possible.

Use this signal only when absolutely necessary.

Wave your arms overhead while holding a helmet, lifejacket, or other object to emphasize the motion. Use the same motion with a paddle. If out of sight, give 3 long blasts on a whistle.

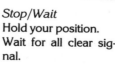

Stop/Wait
Hold your position.
Wait for all clear signal.

Form a horizontal line with your arms and move them up and down. Use the same motion with a paddle. Your arms should resemble the lowered crossbar at a railroad crossing.

Signal   **Execution**

All Clear/Run Center
Proceed ahead.

Form a vertical line with one
arm above your head. The paddle should be held in a vertical
position with the flat part of the
blade visible. Your arm represents the raised crossbar at a RR
crossing.

Run Right/Left
Indicate the best route.

Extend your arm at a 45° angle
in the direction of the preferred
direction. *Always point towards
a safe route; never point to
danger.*

Its important that signals be repeated by each person who sees them in
order to acknowledge receipt of the message and to pass the word on to those
who are following.

This system has been tested and found workable by several clubs.
Because a universal set of signals will aid paddlers from different areas to
paddle together with less confusion, these signals will be part of the AWA
Safety Code. This doesn't mean that local clubs can't add new signals to cover
other situations. Everyone should, however, be aware of the standard signals.

Zoar Gap on the Deerfield River — Debby Arnold

Part Two

Introduction

The descriptions in this section give the paddler reasonably detailed accounts of particular trips. In most cases, descriptions are based on notes taken while paddling the trips, although a few are based on scouting notes only. The rivers themselves receive primary attention rather than the surrounding countryside. The descriptions all follow the same basic format. Key facts associated with the trip appear at the top of the page for easy reference. Key facts are also listed in the table on pages 8-13.

The first paragraph gives an overview of the river and sets the character of the run. Subsequent material details the joys and difficulties to be encountered. Events are described sequentially.

Distances from one spot to the next are not always given. It may be necessary to paddle some distance before the next significant location comes into view, even though it is the next sentence or paragraph in the text. The final paragraph discusses the gage situation for a river, and gives the exact location of the appropriate gage if it exists. The last paragraph also lists names of other nearby rivers that have a similar difficulty rating to the one described, and which could provide some variety for a second trip of the day or weekend.

As a rule, each description is based on a single river level. This level is usually indicated. There are also frequent comments about the river at other levels. If a river is described at a low level and you find it at a high level, its character can be completely different. The opposite is also true: even the most pugnacious rivers look like pussycats when they are void of water.

Distances, when used to describe rapids, are approximate, not exact. For instance, the real business part of the Staircase rapids on the Swift River is roughly twenty yards long, as opposed to fifty yards. Most measurements are the result of eyeballing. Landmarks are noted, sometimes when no difficulties are near, so the boaters may have a fair idea where they are. Usually only those landmarks that seem permanent are listed: houses, concrete walls, etc. Downed trees are considered temporary and therefore are not noted.

The directions left and right are always with respect to a paddler facing downstream.

The reader should realize that a river and its surroundings will change with time due to natural or artificial causes, so do not take the following descriptions as gospel. Think and observe on your own. Great effort has been made to insure accuracy, but it cannot be guaranteed.

Remember, for the level ratings, only the highest reading in an interval is given for the TOO LOW, LOW, MEDIUM, and HIGH levels. For the TOO HIGH level, the lowest reported value is given.

Finally, it bears repeating that this is a guidebook, not a ticket to enjoyable and safe boating. Your success and pleasure on a trip depend primarily on you, the boater, on your paddling skills, your ability to cope with unexpected problems, and on your judgment. No guidebook can ever hope to substitute for these. Guidebooks can only take away a little of the uncertainty.

GOOD BOATING!

Wild and Scenic Rivers

In 1968 Congress passed the National Wild and Scenic Rivers Act. Its purpose was to preserve certain rivers and sections of rivers in their natural free-flowing state and to offer protection to some of the country's most outstanding waterways. The Act was also intended to complement the established policy of dam and other construction on rivers so every river would not be covered with concrete. The scope of the Act is rather far-reaching in that it affects not only the river itself, but also its shoreline. The Act was not intended to restore rivers to their previous wilderness state, but merely to protect them as they exist now.

The rivers to be covered under this Act should meet some or all of the following criteria:

1.   Be a free-flowing river or stream;
2.   Be free of certain types of alterations like dikes, levees, channelizations , dams, etc.;
3.   Have a largely undeveloped shoreline;
4.   Be at least 25 miles in length; and,
5.   Have outstanding scenic, recreational, geological, historical, or cultural value.

Under this system, rivers will be placed into one of three categories:

Wild River — Rivers that are free of impoundments, whose shores are inaccessible except by trail, and whose watersheds are unspoiled and primitive.

Scenic River — Rivers that are free of impoundments, whose shores are largely primitive and undeveloped but accessible by road in a few places.

Recreational River — Rivers that are easily accessible by road, whose shorelines are somewhat developed, and whose waters may be impounded in places.

The rivers that attain the status of a national river will be managed either by the Department of Agriculture, by the Department of the Interior, or by a comparable state agency if the river is in a state protection plan. A river can be placed into the system either by an act of Congress or by designation of the Secretary of the Interior if the river is already in a state system. In almost any case, a detailed study of the river and its watershed must be made prior to a river's introduction or enactment as a wild or scenic river. When Congress first passed this legislation in 1968, it designated eight rivers for immediate inclusion and provided for a ninth shortly thereafter. Twenty-seven rivers were originally designated for study. Since then, a number of additional rivers have been pinpointed for study. Appendix V lists the rivers in New England that have been identified as meeting the minimum criteria for further study and/or potential inclusion into the national system as of January, 1979.

The reader should note that several rivers on the list in Appendix V are white water streams covered in this book. This legislation has power to preserve these rivers in their present natural state, and so, to insure their availability for future recreational activity. Citizens and lobbying groups can influence a river's gaining wild or scenic status; every boater may participate or not, depending on individual inclination. River runners should at least be aware that this legislation exists, and that it can help to curtail future development and exploitation of rivers. One lobbying group which is keenly interested in keeping our rivers free and flowing is the American Rivers Conservation Council at 317 Pennsylvania Ave., Washington, DC 20003. If you are interested in this aspect of river work, give them a call.

MAPS

Maps of every trip accompany the trip description. They were prepared by tracing U.S.G.S. topographic maps, and they should help to orient the paddler and complement the written descriptions. A glance at one of these maps will show immediately if a trip is north-south, east-west, or something in between, and should also provide an overview of the trip. The maps also indicate the high spots of the trip — for example, starting and stopping points, rapids, dams, waterfalls, gage locations, and other points of interest. Nearby roads are indicated to help in planning shuttles and in getting to and from the river. Because topos were used as the base, not all existing roads are shown, although the majority are. A good road map along with these maps should be all that you need to get there and back.

A map key follows in which all map symbols used are explained. It should also be noted that the north arrow indicates magnetic north. Rivers and water in general are drawn in blue for easy viewing and where appropriate, an arrow shows the direction of the trip. A scale is also included on each map to help you judge distance.

MAP KEY

～～～～～	River
▬▬▭▬▭▬▬	4 Lane highway Limited access
▬▬▬▬▬▬	Federal or State Highway
─────────	Secondary road
==============	Dirt road
─ ─ ─ ─ ─ ─ ─	Trail
+++++++++++++++++	Railroad
──·──·──·──	Powerline
── ──── ──── ────	State line

⇨	River direction
⌒ or	Dam
⚑	Campsite
†	Church
◢	School
▮▪	Buildings
⌒⌒⌒	Mountain

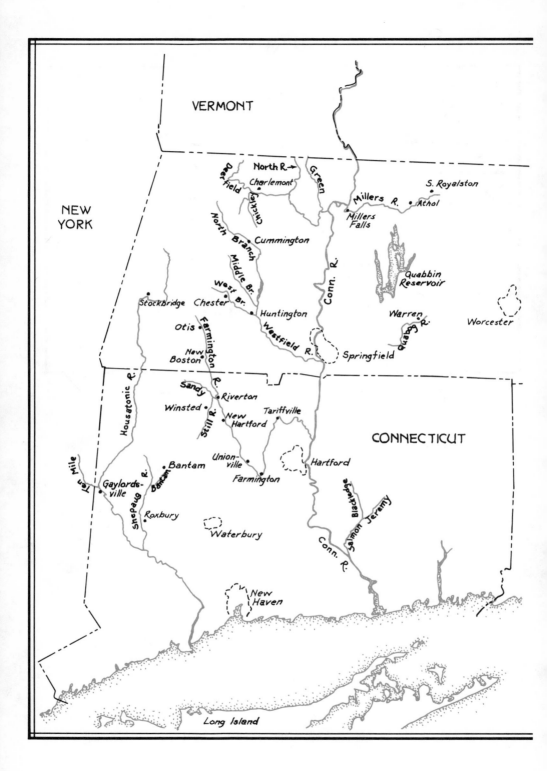

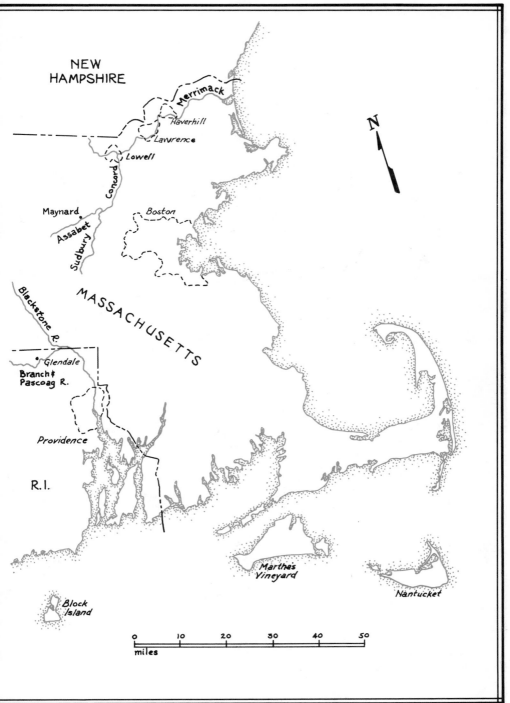

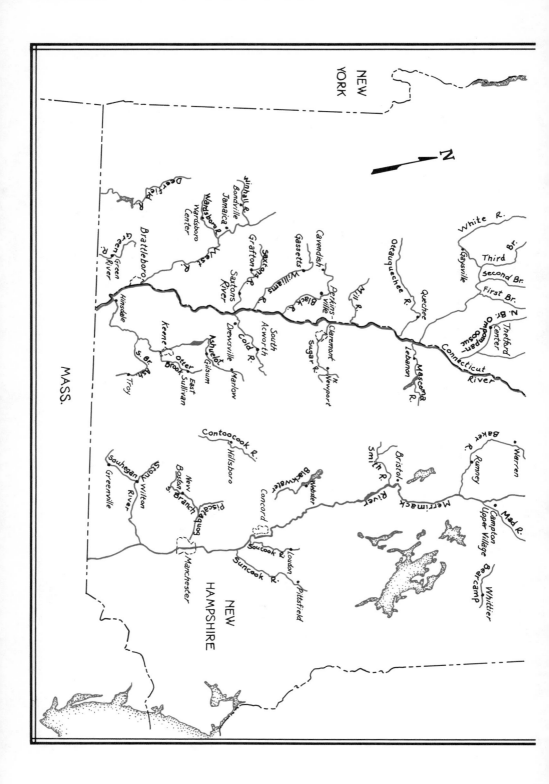

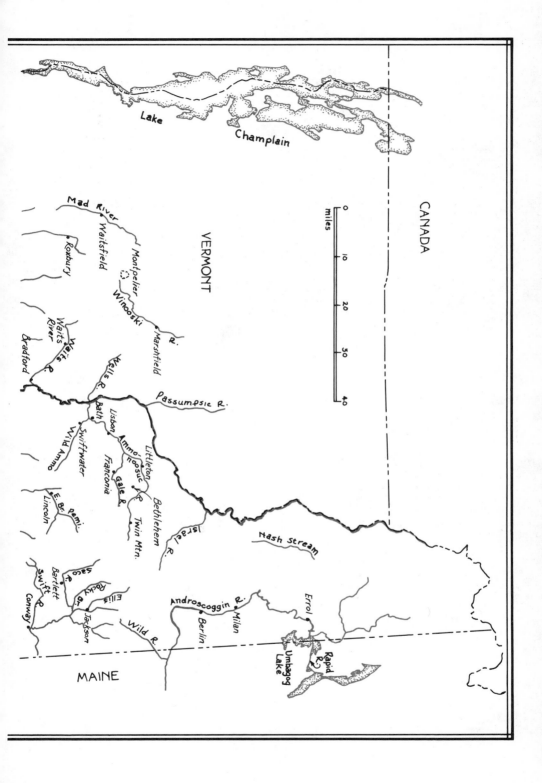

Ammonoosuc River (NH)

RIVER BEND TO PIERCE BRIDGE
Trip A

Distance (miles)	Average Drop (feet/mile)	Maximum Drop (feet/mile)	Difficulty	Scenery
3.0	43	50	2-3	Good

TOO LOW	LOW	MEDIUM	HIGH	TOO HIGH	Gage Location	Shuttle (miles)
	3.6	5.0C	5.0		Bethlehem	3.0
	3.6		5.1		Bath	

Born in the Lakes of the Clouds in the heart of the Presidential Range, the Ammo enjoys a relatively long season, an attractive river valley, and presents an excellent course for training intermediate-level canoeists. In this upper section (Trip A), the rapids are closely spaced and Class 2-3 in difficulty depending on the water level. Only one may require scouting and then only at higher levels. The pools behind the many rocks are just waiting for eddy turns. In comparison, the lower section (Trip B) has much Class 1-2 water, but it is also laced with several Class 4 rapids. The lower half is the less attractive of the two. As with the other White Mountain rivers, the Ammonoosuc can rise rapidly during a heavy rain or a fast melt, so it is not uncommon for it to change levels by a foot or more in an hour. Both the upper and lower trips can be covered in a day's time.

An open canoe below Alder Brook Rapids on the Ammonoosuc
— Gordon Calverley

To start a trip, put in at Twin Mountain where Route 3 crosses the river or, what is usually done, some 2.2 miles downstream where the Ammonoosuc comes close to the road. A motel is located here (River Bend). The two miles down from Twin Mountain are mostly Class 1-2 and they are good for warmup or to lengthen the trip. At River Bend, the river is some 50 to 75 feet wide, with a sprinkling of small rocks that are mostly covered in medium water. Below River Bend, the Ammo turns left, flows away from the road, and the rapids begin. At first there is 75 to 100 yards of easy Class 3 water pouring over and around rocks. A sharp right turn with a boulder on the outside follows. Another playful Class 3 rapids comes shortly; there is a pause, then another left turn. A pool follows the last rapids, then there's a chute where rocks extend out from the left bank at the top. Here some waves are 2 to 3 feet high at gage readings of 3.6 to 4.8 (Bethlehem). These rapids typify the rest of the run, although the pace will soon let up slightly. Except for very high water, there are always rocks making the path somewhat less than straight. Up until the end of the trip, there should be no special difficulties.

Boat Breaker Rapids is the last and the hardest rapids on this trip. Near the end of the trip, notice a small island on the left, and below, a slight right turn with some houses on the left bank. Boat Breaker Rapids is 50 to 75 yards in length. It starts with a series of small abrupt drops among large boulders, moves into a fast channel of turbulent water, and finally finishes with another small, abrupt drop with a covered, or partially covered, rock directly in the middle. The main channel starts center, shifts slightly left, then splits around that rock in the middle. In low or medium water, it is a straight run, and there is plenty of room. At a gage reading of 4.5 or higher, Boat Breaker Rapids will swamp most doubly paddled open boats if tackled directly. Closed boats should have no trouble at these levels. If the river is high, and you are in doubt, look these rapids over. At a gage reading of 3.6 (Bethlehem), Boat Breaker is rated Class 3; at 4.5 it's a 3+; at a reading of 5.0, it is a Class 4. Below, on the left bank, is the government gage. Following the gage there is an easier rocky rapids leading to Pierce Bridge, where the take-out is on the left, up a steep sandy bank. Muchmore Road forks off Route 302 and parallels the river up to Boat Breaker Rapids on the left side. The HIGH rating for open boats is mainly for Boat Breaker Rapids, since the rest are not so heavy, although they do deserve respect.

When paddling the Ammonoosuc, or any other New Hampshire river, be aware that state law requires cars to be parked entirely off the road. For this area, group camping can usually be carried out at Zealand Campsite which is upstream from Twin Mountain.

There are two gages on the Ammonoosuc. The upstream Bethlehem gage is in Grafton County, on the left bank, 0.2 miles upstream from Pierce Bridge on Muchmore Rd. The other gage is also in Grafton County, on the left bank, 0.4 miles downstream from the Wild Ammonoosuc River, and 1.5 miles

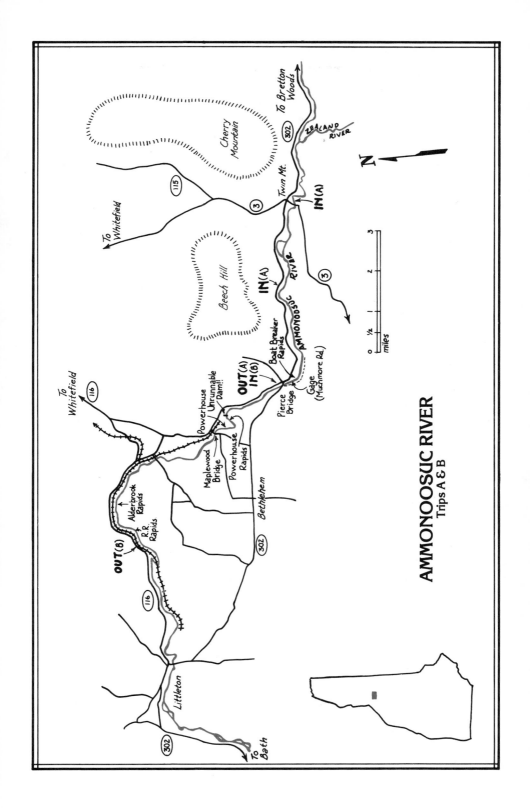

AMMONOOSUC RIVER
Trips A & B

downstream from Bath, New Hampshire. Both gages have outside markers although the Bethlehem gage is sometimes hard to read. The Bath gage can be read remotely by the U.S.G.S. Another river in the area that is of similar difficulty is the Wild Ammonoosuc.

Ammonoosuc River (NH)

PIERCE BRIDGE TO ROUTE 116
Trip B

Distance (miles)	Average Drop (feet/mile)	Maximum Drop (feet/mile)	Difficulty	Scenery
7.0	35	55	1-2-3-4	Fair

TOO LOW	LOW	MEDIUM	HIGH	TOO HIGH	Gage Location	Shuttle (miles)
		4.5C	4.5C		Bethlehem	7.0
		5.0C	5.0S		Bath	

The lower Ammonoosuc (Trip B) offers a longer run than the upper (Trip A). It has two Class 4 rapids, much easy water, and a portage around a 20-foot dam. The scenery is not so attractive as Trip A above because of the signs of encroaching civilization that populate the banks. The first mile below the dam can usually be run even when other nearby streams are dry—it may be only Class 2, but at least it's wet.

Putting in at Pierce Bridge, or continuing down from the upper run on the Ammo, it is slightly less than two miles to an abandoned 20-foot dam and powerhouse. For most of the way there is a current with some easy Class 2 rapids. The backwater from the dam starts just upstream before a left turn that brings the paddler to the dam itself. The portage is best on the right, over a concrete wall, and down a short hill. This carry is a real pain in the keel, but it is absolutely necessary. You can look at it as good portage practice. Below the dam is approximately one mile of continuous rapids.

The initial 100 to 150 yards below the dam are Class 2-3. The Ammo then turns right and then in a short distance it turns left, entering into Powerhouse Rapids (Class 4 at medium or high levels). At this spot, there is a large island on the right side with a very narrow channel to its right that exists only in high water. Narrowed by the island, the Ammo races down a 100-yard chute on the left side. Halfway down is a collection of rocks in the middle and on the left. At a gage reading of 4.5 (Bethlehem), these rocks are mostly covered and the entire course is an angry, churning broth, full of holes and hydraulics hidden by haystacks. The right center channel is usually taken at this level. Once it's set up, it's a straight shot. In lower water, more maneuvering is required to avoid the menagerie of rocks that appears. To scout, land at the upstream end of the island or walk down from the dam. Just before the left turn entrance into Powerhouse Rapids, there are some boulders on the left. After Powerhouse Rapids, fast calmer water rushes the paddler to a sharp right turn where there's an abrupt river-wide drop extending from the left shore. Most severe on the left and in the center (2-foot drop in high water), it can be skirted on the extreme right, but it's difficult to get there because the current pushes you constantly to the outside left. The associated hydraulic is a honey — get too close and it's sticky. The river then continues Class 2-3 past another old powerhouse on the right shore, pauses long enough to catch a breath, then drops into another good rapids rated Class 3 or 3+ in high water. At low levels, be careful of the many sharp rocks here. After another left turn, there are easier rapids leading to the Maplewood Bridge.

Below the Maplewood Bridge some Class 3 rapids start out easy, but three-quarters of the way down, a rock ledge pushes out from the left (1.5-foot drop), as does a smaller one near the end, from the right. Several easier Class 2-3 rapids follow. For the next eventful one, notice several large rocks, especially one on the right. Here there is a chute with, in high water, another of those stairstep descents on the left tapering down a bit on the right. A stretch of Class 1-2 water follows as you pass a dead-end road on the left and an iron bridge. Below this bridge are two islands with runnable channels on either side. On the right, Route 116 approaches close for the first time as the Ammo continues on with more Class 1-2 water. This section of calmer water ends with a 1.5-foot drop over a rock ledge, just upstream from a pink, box-like house on the right bank. For several hundred yards afterwards, nothing much happens; then one enters Alder Brook Rapids.

Lying in wait just around a slight left turn, Alder Brook Rapids is heavy, fast, and rocky — Class 4. It is sometimes recognized at its start by a rock on the left whose top contour resembles a camel, but by the time you spot this figment of someone's imagination, you're into it, so just listen for the sound of rushing water. Since a line of rocks extends out from the right side, the paddler is forced to begin left, or left center. From here, one usually proceeds down the

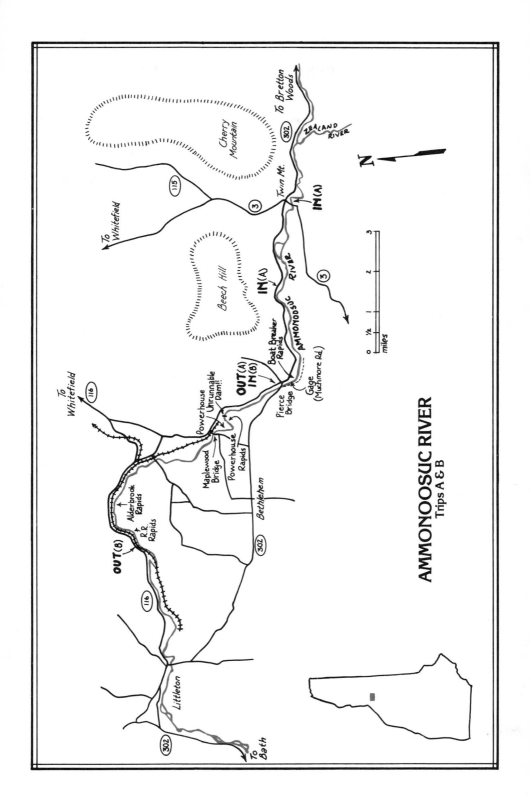

AMMONOOSUC RIVER
Trips A & B

center with the main current and then either left, or right, around a boulder in the middle three-quarters of the way down. This boulder usually supports a strong upstream pillow and most paddlers choose left. Alder Brook lasts 25 to 50 yards, and has some nice holes and haystacks measuring 2 to 3 feet in high water; it can also be run straight on the left side.

Ten minutes' paddle from Alder Brook brings you to Railroad Rapids and the trip's end. Starting in a right turn as the current picks up, rocks flash by like railroad ties, and then shortly the Ammo turns sharply left, falling over more rocks with the turbulence prominent, but less than at Alder Brook. Near the bottom is a 1.5-foot drop over a rock ledge on the left and center. In general, the whole left side is cluttered with rocks. The right side is a series of small drops and is the preferred route. The take-out is on the right bank where the rocks are extremely sharp and the bank is steep. A short walk up the embankment leads to a railroad track and an old dirt road where the drivers were supposed to meet you.

The canoeability ratings of this trip are with respect to the heaviest rapids only, as the rest can be easily run by an open boat. Also, if skilled, a solo paddler can even negotiate an open canoe on these harder ones at LOW or MEDIUM levels.

There are two government gages on the Ammonoosuc. Closest to the mouth, the gage near Bath is in Grafton County on the left bank, 0.4 miles downstream from the Wild Ammonoosuc River and 1.5 miles downstream from Bath, New Hampshire. The concrete gage house is clearly visible from the road (Route 302). The Bath gage is in the Telemark system. The Bethlehem gage is also in Grafton County, on the left bank, 0.2 miles upstream from Pierce Bridge on Muchmore Road and 3 miles east of Bethlehem.

Androscoggin River (NH)

ERROL TO PONTOOK SCHOOL

Distance (miles)	Average Drop (feet/mile)	Maximum Drop (feet/mile)	Difficulty	Scenery
20	4	14	1-2	Good

TOO LOW	LOW	MEDIUM	HIGH	TOO HIGH	Gage Location	Shuttle (miles)
	1500 CFS				Errol Dam	20

The Androscoggin is not a white water river in the classic sense, yet it does have one characteristic that makes it worthy of description: it runs in the summer. When most other rivers are dry, the Androscoggin keeps chugging away, even into August and September. The rapids are not noteworthy, but there are some to be found; the scenery isn't super, but it's certainly enjoyable; the drive to the river isn't short, but it's bearable. For beginning open boaters and hard-core racers who need a river fix, the Androscoggin is one of the best opportunities around when the sun is high and the snow is long gone. The rapids are clustered into just two or three spots, and the remaining part of the river moves about as fast as the federal bureaucracy.

Passage to the Androscoggin usually goes through Berlin, New Hampshire, which is dominated by the Brown Paper Company. The smell is evident

10 miles downwind, and the city is under a blanket of belching factory exhaust. However, without Brown, you probably would not have a summer run on the Androscoggin, or on the Rapid, for that matter. The dam at Errol which controls the river flow does so in response to factors in Berlin. At any rate, the Androscoggin is there, so why not use it. The upper parts described here are not really objectionable.

The dam at Errol has several sluiceways to regulate the river's flow. Located about 0.8 miles north of Errol, it is reached via a small dirt road off Route 16. As the water comes blasting down the spillways, strong eddies are created on the sides, and a group of standing waves sits in midstream. The rapids continue several hundred yards down to a large pool. This pool, visible from Route 16, is one spot from which a trip may be started. A U.S.G.S. gage is in this area, but it has no outside staff. An alternative starting spot is on the right side, upstream from the Route 26 bridge in Errol. Here there is a small road that runs parallel to the river and alongside the heaviest rapids of the trip. Saco Bound has a small store and canoe rental agency on the road. At typical summer level the dam releases about 1500 CFS which corresponds to two to three spillways being used.

The rapids under the Route 26 bridge in Errol are the heaviest of the trip. They begin just after a slight right-hand turn from the pool above and proceed for several hundred yards. At 1500 CFS, standing waves measuring up to 2 feet (trough to crest) are the biggest obstacle; the difficulty is heavy water, Class 2. A line of rocks extends from the right bank at the start, and the current is quite swift, but could be neutralized by two strong paddlers. A run down the middle should be straightforward. The river is over 100 feet wide here and these rapids end in a pool-like expanse.

Thirteen Mile Woods begins just outside Errol on Route 16. The river is flat here and continues to be for about 3.5 to 4.0 miles. After this distance there is a series of small rapids interspaced with calmer water. These rapids continue for about 1 to 1.5 miles. All are easy Class 1-2 in difficulty; waves may be as high as one foot; and there are numerous very small drops over ledges and rocks.

A small red bridge is the next landmark. The current speeds up under the bridge and seven islands are located on the right, downstream side. Thirteen Mile Woods ends in another 2 miles.

When the landscape becomes marshy, you are approaching Pontook Dam. Another signal is the appearance of a small artificial stone island. Pontook Dam must be portaged, and the carry is not easy. The dam is made up of a collection of logs, brush, and stones all piled in a jumble. Route 16 can be seen from the river when you are upstream from the dam. Downstream, there is a collection of Class 1-2 rapids and a fairly consistent current. When Route 16 comes close again, be prepared for the best rapids of the lower half of the trip. There's lots of white water caused by ½-foot drops over ledges and rocks. The paddler is forced to plan and choose a route for several hundred

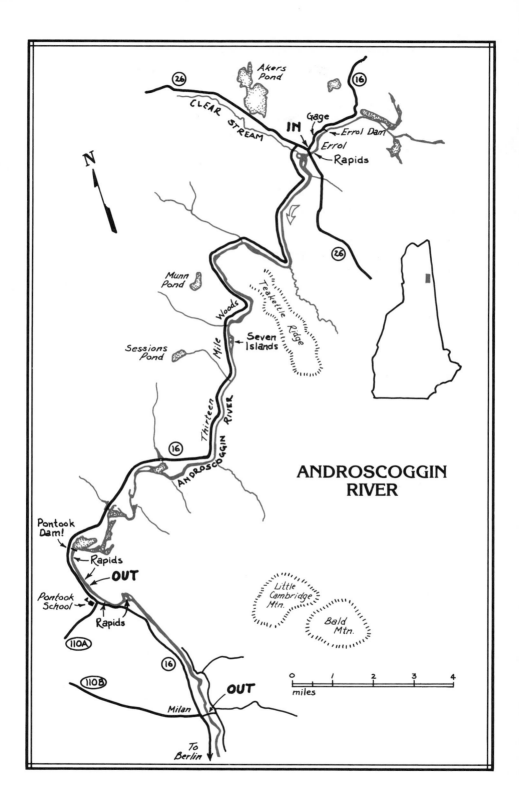

yards (Class 2 at 1500 CFS), but the current is not so powerful as it is in Errol. These interspersed rapids continue for some 2 miles from Pontook Dam, but the price of admission for running them all is the absence of a convenient take-out. It is probably best to take out where Route 16 comes close to the river below the dam unless you want an extended trip. There are several roadside turnoffs in this area. After the rapids, the river meanders lazily through pasture lands to Milan (4 miles). Although this portion is flat, the distant scenery, the White Mountains, is rather pleasant.

Pontook Dam can be reached via a narrow, unmarked, U-shaped dirt road off Route 16. Drive to the end of the U, and walk the rest of the way. This road cannot be seen from the river. A small piece of water lies on the west side of Route 16 just north of this dirt road.

There is no convenient gage for the Androscoggin. For river information, try calling Saco Bound/Northern Waters in Errol. Be aware that the local residents are somewhat skeptical of outsiders, so don't give them any reason to restrict river access and travel. Note also that permits are required for camping in the area.

Assabet River (MA)

ROUTE 117 TO ROUTE 62

Distance (miles)	Average Drop (feet/mile)	Maximum Drop (feet/mile)	Difficulty	Scenery
1.5	10		1-2	Unfortunate

TOO LOW	LOW	MEDIUM	HIGH	TOO HIGH	Gage Location	Shuttle (miles)
2.5	3.0				Maynard	1.5

Traditionally, the Assabet is among the first rivers to be paddled in eastern Massachusetts, especially if you are just entering white water. Small, slow-flowing, with well defined rapids and calm water alternating throughout the run, the Assabet is ideal for easy Class 2 boating except for one slight drawback; the river is one of the least scenic and biggest eyesores around. You will find all kinds of interesting ingredients making up this cauldron: dead fish, arm chairs, old refrigerators, stoves, shopping carts, and even some slightly used cars. And, to top all this, the river smells. Other than that, the Assabet is a great place for boating. If you must go there, consider yourself warned.

The trip itself is very short, so you can repeat it several times in one day, that is, if your nose and stomach can take it. Start the trip just above where Routes 62 and 117 intersect on the west side of town. There is a small dirt road here that runs along the right side of the river, next to a box-like building (Pace's). Above the bridge there, the current is gentle, and there's a ½- to 1-foot drop immediately past the bridge. This drop is more abrupt on the right side than on the left, although in medium or high water this difference is not noticeable. There is a dam several hundred yards upstream of this bridge that ought to be opened to flush out the river.

The approach to the next bridge has one of the more difficult rapids on the trip. The current speeds up, hobbling over small rocks in the process as the water gathers momentum and runs directly into the middle bridge support. The boater can go to either side, although the right side offers a better outrun. Directly downstream of this bridge, there are several small islands. This latter section is scratchy at a gage reading of 2.7. Next is a series of Class 1-2 rapids barely noticeable in low water. Another concrete bridge then appears.

As the Assabet approaches Maynard Center, the river is walled in on both sides and the current stiffens a bit. There is a Digital Equipment Corporation plant on the right. This section ends as the river turns left and passes under a bridge. For several hundred yards beyond this bridge, there are uninterrupted Class 1-2 rapids.

Just upstream from the next bridge (Route 27), there is a river-wide ledge that drops about 1 to 2 feet. The gage house is on the right bank, immediately upsteam from this drop. The ledge itself can be run almost anywhere, but be aware that there are rocks in it. From here to the take-out you should find nothing new.

Take out near Alphonse's Restaurant parking lot, which is on Route 62 east of Maynard Center. As with the put-in, you should ask permission before using private property for hauling canoes and changing into or out of your wetsuit.

The gage is located in Maynard, on the right side, 75 feet upstream from the Route 27 river crossing, almost directly across the street from a local French restaurant.

For another river of similar difficulty, try the Piscataquog in New Hampshire. It's farther away, but it will not dissolve your fiberglass boat.

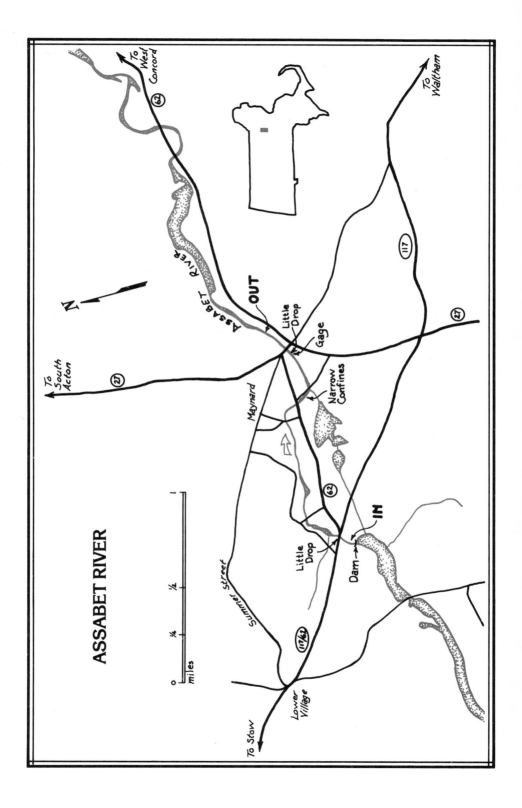

ASSABET RIVER

Ashuelot River (NH)

MARLOW TO GILSUM GORGE
Trip A

Distance (miles)	Average Drop (feet/mile)	Maximum Drop (feet/mile)	Difficulty	Scenery
3.5	63	80	3-4	Good

TOO LOW	LOW	MEDIUM	HIGH	TOO HIGH	Gage Location	Shuttle (miles)
3.6	4.1	5.5C	7.5C		Gilsum Gorge	3.5

The Ashuelot has its origin in several lakes to the north of Marlow, New Hampshire; it flows generally southeastward to meet the Connecticut below Hinsdale. Along most of its course the Ashuelot is flat, although two sections do offer good white water challenges, and another is an appropriate training ground for beginners. The upper Ashuelot (Trip A) is small and requires precise boat control in maneuvering through the many rock patterns. There are at least a half-dozen of these goodies, plus one that is substantially harder. In low water the passages are tight. Higher water opens new routes, but packs them with lots of turbulence. If the gage reading is over 5.0, the rapids tend to blend together forming one long, difficult Class 4 rapids. At the trip's end, there is an impressive gorge. Trip A is short, and can be easily repeated in a day. A road follows the river closely, which greatly facilitates the shuttle.

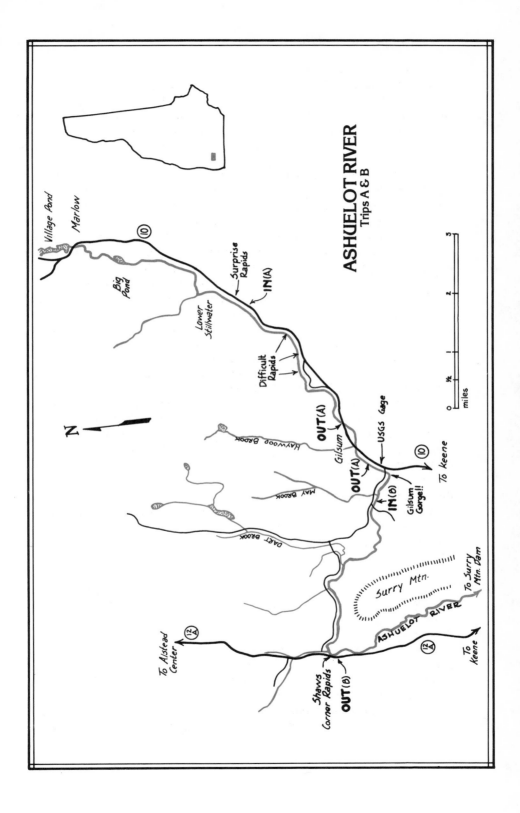

ASHUELOT RIVER
Trips A & B

Village Pond
Marlow
10
Big Pond
Lower Stillwater
Surprise Rapids
IN (A)
Difficult Rapids
N
Haywood Brook
Gilsum
OUT (A)
OUT (A)
USGS Gage
MAY BROOK
IN (B)
Gilsum Gorge!!
10
To Keene
DART BROOK
Surry Mtn.
To Surry Mtn. Dam
ASHUELOT RIVER
12
A
To Keene
12
A
To Alstead Center
Shaws Corner Rapids
OUT (B)

miles
0   ½   2   5

There is no single spot to start the trip on the upper Ashuelot, although the recommended put-in is by an old section of Route 10, some 3 miles north of Gilsum. A quarter-mile above there is a small bridge crossing the river. Above the bridge the water is flat. If one is paddling down from this flat section, Surprise Rapids (Class 4) is first. Not normally a rapids to start out on, Surprise bears study for a possible later run with warmed-up muscles — or you may decide to forego the whole thing.

In low water Surprise Rapids appears as follows. As you approach the rapids the current speeds up and the left side is choked by a boulder, forming a chute in the right center that drops 1 to 2 feet into some haystacks. After a short stretch of fast, calmer water, several rocks, sitting in the middle, funnel the current to either side. The left side has an abrupt drop into a hydraulic and a series of standing waves in a narrow, but straight, outrun. The right side has a gradual drop, but a rock in the middle makes things difficult. Following almost immediately are several more drops, lots of turbulence, and finally a 2- to 3-foot drop at the end, with a strong hydraulic afterwards. The entire length of Surprise is about 100 yards, the current is very fast, and the channel is narrow. Surprise is the hardest rapids on the run, and the usual starting point is below it. Surprise Rapids is visible from the road and should be scouted prior to running every time. Large rocks line the banks, and there is no easy way. If the gage is reading 5.0 or over, Surprise is really mean, even to the competent, and the last hydraulic looks like a keeper (it isn't always).

After such an outburst, the Ashuelot quickly composes itself with a steady current, a sprinkling of rocks, and an easy stone's throw width. The rapids in this section are Class 2-3 depending on water level. Where an island divides the river, the right side is sportier than the left. There are several rocks on the outside of the turn, at the downstream end, just as the other channel returns its water to create a crisscross of currents. As the road comes close on the left side, the water becomes calm, forming a small pool-like area as the river prepares for a 50-yard narrow chute. This rapids marks the beginning of the more difficult water. From here to Gilsum, rapids are fairly continuous. At a gage reading of 4.1, they are technical Class 3; at a gage of 5.5, they are strong Class 4.

In the rapids last mentioned above, the current speeds up and standing waves sprout as the paddler races toward several rocks at the end. Placed inconveniently in the middle, these rocks force the boater to go to either side. If these rapids give trouble, it is best to pull out, as the rest are harder. Below there is a chance for a brief rest, then the discharge filters between two large boulders on either side as the river prepares for a short S-turn with a small drop at its end. In a short while, the current picks up again, the river turns left, and larger rocks clutter the turn, while rocks below generally clog the river's path. In the turn itself, there is an abrupt drop.

After a right turn there is a very fast chute with extremely turbulent water at high levels. The chute lasts down to, and around, the next left. Rocks in the center halfway down force a large portion of the current to the left, then sharply right in a tight S. In medium or high water, it is possible to go through the rocks, although there are souse holes on the downstream side. This is probably the hardest rapids normally run, except for Surprise.

Following the next left turn are more large rocks with an abrupt drop near a flat rock in the center. Left center is clearest, but drops the paddler into a good hydraulic. More haystacks follow.

After another left turn another chute begins with several drops and more rocks. There is a house on the left bank, slightly downstream from the start.

The Ashuelot then turns left again into two more drops that are easier than those above. The remains of an old bridge can be seen on either bank, and a small dirt road on the left goes up a hill to meet Route 10. This marks the end of the most difficult rapids, although plenty of Class 2-3 water remains before Gilsum.

Houses soon appear on the left bank, and an island divides the river. The left channel is blocked by the more resilient remains of a dam while the right side can be run almost anywhere over the 1- to 2-foot drop. From here it is only a short distance to the bridge at Gilsum, where one can take out by a lumber yard on the left.

It is possible to extend this trip to the next bridge (1 mile) where the rapids, although still Class 2-3, are not so difficult as above. There is one particularly deceitful rapids just upstream of and under this bridge. In high water, these rapids shouldn't be underestimated, as they are tough. The take-out for this extension is below the bridge, around a left turn. There is a roadside turnoff near this exit. Extreme care should be taken in exiting the river, especially in high water or with a weak group, because Gilsum Gorge is several hundred yards downstream. A swim started at the take-out could sweep a person through the gorge — definitely not recommended.

Gilsum Gorge has been described as being absolutely unrunnable. It is not really unrunnable, as are the gorges on the Black, the Smith, or the Cold, although it is difficult. When scouted by land, it appears that only several maneuvers would have to be made, but they are real winners, and there would be little leeway for mistakes, especially as the water level rises. The banks are extremely steep, limiting accessibility to mountain goats and the neighborhood kids. The Gorge has been run by kayak at a gage reading of 4.0. Make your own decisions.

At a gage reading of 7.5, the Ashuelot is runnable, but only for the very experienced, who are familiar with the river, and then only after some scouting to check the location of the bigger souse holes. At this level, the Ashuelot is boiling with rage and screaming down the countryside. A mistake means a long, difficult swim. Perfection in planning and execution is required. Sounds like a Class 5 run, doesn't it?

The gage is located on the right bank, 60 feet upstream from an impressive stone arch bridge guarding the entrance to Gilsum Gorge. This bridge is just off the Keene-Newport Road (Route 10).

Other nearby rivers that are similar in difficulty are the Otter Brook and the South Branch of the Ashuelot.

Dropping into the final hole on Surprise Rapids on the Upper Ashuelot — Dick Siegel

Ashuelot River
(NH)

GILSUM GORGE TO SHAWS CORNER
Trip B

Distance (miles)	Average Drop (feet/mile)	Maximum Drop (feet/mile)	Difficulty	Scenery
4	30	50	2	Good

TOO LOW	LOW	MEDIUM	HIGH	TOO HIGH	Gage Location	Shuttle (miles)
	4.6	5.7	6.5	7.5	Gilsum Gorge	4

The middle Ashuelot (Trip *B*) is a fine Class 2 run for instructing beginners: the rapids consist mainly of small haystacks, the curves are gentle, yet it takes some skill to negotiate them, and there is a tough Class 3 rapids to end the trip. A road follows alongside most of the way, although for the majority of the trip the paddler is unaware of it. The unfortunate aspect of this trip is that it seldom has enough water for an enjoyable run. It should also be mentioned that this trip seems straightforward even when the water is high, but coping with the strong current may be too much for people with only Class 2 ability. Neophytes should be aware that rescues under these conditions are extremely demanding and difficult, and that boat and boater can be at the mercy of the river.

Start the trip downstream of Gilsum Gorge either where the Surry Mountain Dam Road first comes close to the river or by a small bridge a little farther downstream. The river is fairly wide here, with Class 2 water and a scattering of small rocks. These rapids are typical of what lies ahead. The Ashuelot continues in this way, alternating between Class 1 and Class 2 in difficulty, reaching Class 3 if the gage is above 6.5.

At one spot, just as the river approaches the road on the right bank, there's a little chute and shortly thereafter, the Ashuelot turns right into a long section of steeper gradient. Here the waves can rise to 2 feet at a gage reading of 5.5, and it is necessary to maneuver around small rocks. Little hydraulics also give the beginner a challenge.

At a point below, the main channel funnels to the right side and into a tight S curve with rocks lining the course. The boater must turn first left and then quickly right again as the channel widens. The left side is blocked by rocks and the current in the S is faster than normal. This rapids, rated 2+ in medium water, is the second hardest of the trip. The road can be seen from the river on the right bank. Following this rapids is more Class 1-2 water.

When you see a left turn ahead, with stone blocks from an old building on the right bank, get ready for Shaws Corner Rapids (Class 3). Starting with a small ledge on the inside left of the turn, the current then moves to the outside and subsequently speeds up to form a series of haystacks that line the main channel on the right after the turn. These haystacks end as the water drops over two ledges, about a boat's length apart. A hydraulic follows each ledge, with the wave on the downstream side being some 2 to 3 feet high in medium water. A path to the extreme left easily avoids the ledges. The normal take-out spot is a small footbridge immediately below. Shaws Corner Rapids lasts about 50 yards from start to finish.

The gage is located on the right bank, 60 feet upstream from a large stone bridge guarding the entrance to Gilsum Gorge. This bridge is just off the Keene-Newport Road (Route 10). At a gage reading of 7.5, the Ashuelot is blowing down the mountainside, and the run looks very inviting. Be aware that at this level the current is very strong and continuous: those with Class 2 ability should think about paddling somewhere else. This is the basis for the TOO HIGH rating. If you want another river of Class 2 difficulty, try the Cold.

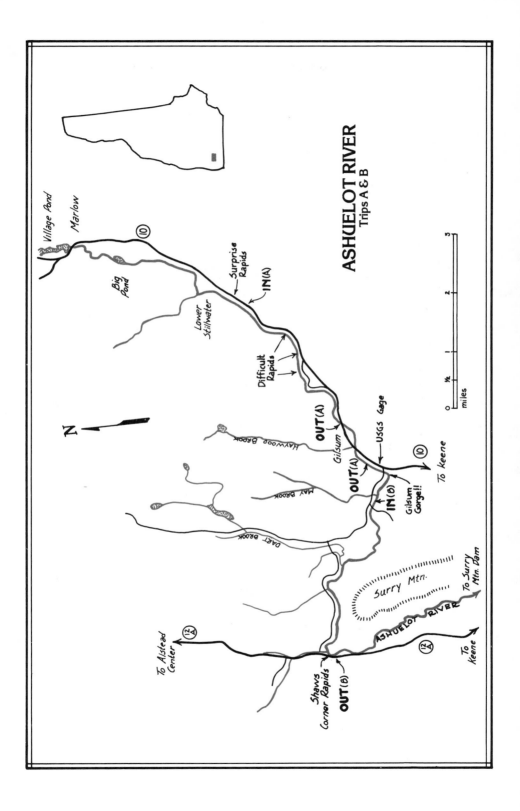

ASHUELOT RIVER
Trips A & B

Village Pond
Marlow
Big Pond
Lower Stillwater
Surprise Rapids
IN (A)
Difficult Rapids
OUT (A)
Gilsum
OUT (A)
HAYWOOD BROOK
MAY BROOK
USGS Gage
IN (B)
Gilsum Gorge!!
To Keene
DART BROOK
Surry Mtn.
To Surry Mtn. Dam
ASHUELOT RIVER
To Keene
To Alstead Center
Shaws Corner Rapids
OUT (B)

N

miles
0   ½   1   2   3

Ashuelot River (NH)

ASHUELOT TO HINSDALE
Trip C

Distance (miles)	Average Drop (feet/mile)	Maximum Drop (feet/mile)	Difficulty	Scenery
3.5	52	80	3-4	Poor

TOO LOW	LOW	MEDIUM	HIGH	TOO HIGH	Gage Location	Shuttle (miles)
		4.8C	5.3C		Hinsdale	3.5

The lower Ashuelot (Trip C) is quite different from the upper sections. This lower section is much wider, has a larger discharge, and is polluted. If portaging dams and pollution are your thing, this is the place to be. In 3½ miles, there are 4 to 5 dams, depending on where you stop. In between dams, the water can be powerful, with one set of especially heavy rapids. Even with its drawbacks, this section of the Ashuelot offers good practice on one of New England's larger white water river sections.

Put in at a covered bridge in Ashuelot, New Hampshire, where the river is wide (it would be hard to throw a stone across) and there are rocks sparsely scattered about. Below, the Ashuelot turns right where easy ripples are characteristic for a while. Within a half-mile the river enters a large pool; the first dam follows. Both sides present tough portages since the water pours through

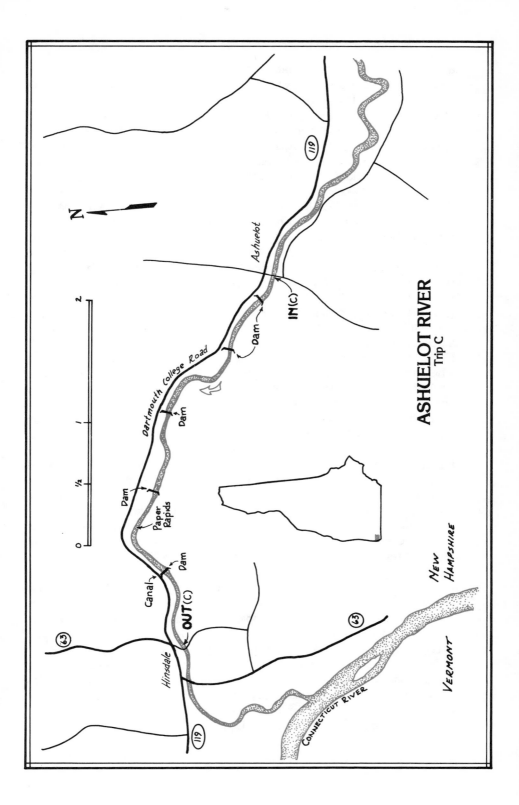

ASHUELOT RIVER
Trip C

a break in the middle into large, powerful standing waves at the bottom. Depending on the level, it may be possible to climb onto the dam itself and lift over from there. There are strong eddies below on either side, and there are also easier rapids on the right side. The river has a current for a while, then it comes to another pool; the second dam follows. Below, the river picks up speed in some Class 3 rapids, then turns right where you'll encounter a good series of bumps, some clear water, and another set of waves at the bottom. Here, a large boulder sits in the middle. After a right turn, there's a long set of standing waves — straightforward but, since the river is large, waves with a lot of power. They continue for several hundred yards; then comes a large pool; the third dam follows. Both sides are difficult for a portage, but the right side is probably best since it avoids the Ashuelot Paper Company on the left bank. The put-in below is also easier on the right. Just below a small bridge there, a little ledge extends out from the right. On the left bank, a house-sized boulder marks the beginning of the heaviest rapids of the trip, Paper Rapids.

Just beyond the third dam, the water is already moving quickly. It gradually picks up more speed and turbulence as it approaches a right turn below the paper company. Here there are very turbulent water and large hay-stacks measuring up to 4 feet at a gage reading of 5.4. The river drops about 5 feet over a distance of 150 yards. The main current funnels into the center where the heaviest water contorts. Strong hydraulics and 1- to 1½-foot drops are common. The approach to Paper Rapids is rock-cluttered, and with the speed and power of the current, it could be tricky. An approach on the right will avoid most of the more troublesome parts. The left side is more difficult because of rocks that extend out from shore. Near the end there is an abrupt drop over a ledge, but it is an easy one compared to those above. With all the aquatic gymnastics, the white water almost appears to be a snow-covered river. A turnoff from Route 119 that passes alongside Paper Rapids provides a good vantage point for scouting, which is probably a good idea. At a gage reading of 5.4, Paper is rated Class 4; at 4.8, it is tough Class 3.

After a left turn the river straightens out in a long rock-studded rapids where a center route is fine. Rocks reach out from the right side. This set of rapids is rated Class 3 mainly because of its length. Below are some strong rapids where the water funnels to the center with 2-foot waves at a gage reading of 5.4. Then the river gets quiet once again, and the fourth dam follows. This dam is somewhat different from the others; the drop over is gradual, and not so large. A few paddlers may, therefore, be tempted to run it. Look it over for yourself. Also, the current continues right up to the dam, so be careful that it doesn't come as a surprise. On the extreme right, upstream side, a shoreline canal taps parts of the river; again, be careful. Directly below the dam, an island divides the channel and either way is fine. The ensuing rapids continue for an extended distance and stop just as the outskirts of Hinsdale come into view. The fifth dam is just below the first bridge, and after this one, you can take out any dam place you desire.

The gage is on the left bank, 40 feet upstream from the downstream bridge in Hinsdale. It does not have an outside staff, but the Corps of Engineers can obtain a gage reading via phone. The canoeability ratings refer mainly to Paper Rapids, although they also indicate what the rest of the run is like.

If you want another river that is similar in size to the lower Ashuelot, try the Contoocook. If you want another Class 4 run, try the upper Ashuelot (Trip A).

South Branch Ashuelot (NH)

TROY TO ROUTE 12

Distance (miles)	Average Drop (feet/mile)	Maximum Drop (feet/mile)	Difficulty	Scenery
2.5	80	100+	3-4	Fair

TOO LOW	LOW	MEDIUM	HIGH	TOO HIGH	Gage Location	Shuttle (miles)
0.0					Route 12	2.5

The South Branch of the Ashuelot typifies a small New Hampshire stream. It is steeply pitched, rocky, flows only in early to mid-spring, and when it is runnable, it is cold. The South Branch is a good complement and close to both the main Ashuelot and Otter Brook. The South Branch is best run when the water is up. In very high water, however, it is a river to be respected. The gradient is steep, especially in the upper portion of the run. In low levels, rocks will annoy the boater who can only paddle in a straight line. In high water, rescues will be very difficult due to the powerful current. Route 12, which follows the river for the entire trip, is convenient for scouting and shuttle.

The South Branch gets its water from several ponds scattered around Troy, New Hampshire. As it flows through Troy, it is very narrow, relatively unexplored, and downed trees could be a big problem. A trip can be started

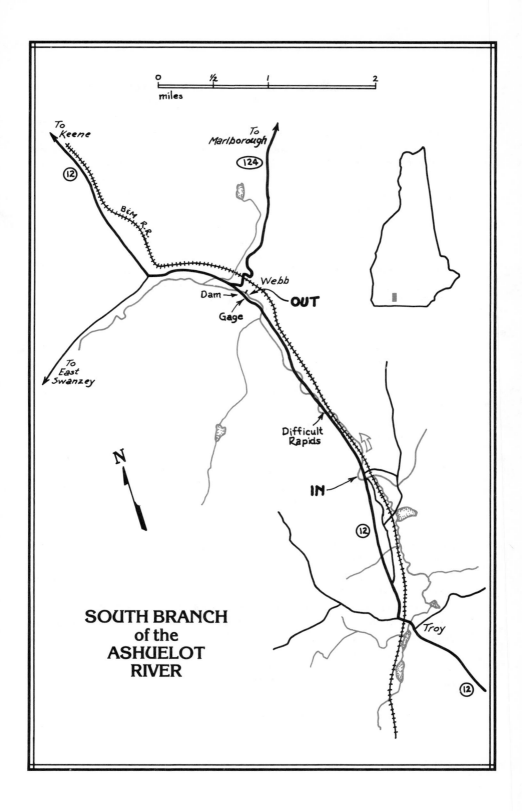

SOUTH BRANCH
of the
ASHUELOT
RIVER

almost anywhere along Route 12 where the river is close. Most people start near a bridge that is about ¼- to ½-mile north of Bowers (a local store). Driving north from Troy, you will approach this bridge while going downhill. The upper portion of the trip is the most difficult, and it allows the paddler little time to warm up. The single hardest section lies in the vicinity of two closely spaced bridges, first a railroad bridge, then shortly downstream a Route 12 bridge. The river makes a sharp left turn here, banking off the right-side RR bridge abutment in the process, which creates some complex currents and standing waves. Once past the highway bridge, the South Branch starts a long sweeping right turn that lasts down to the next Route 12 bridge. This turn has a stone wall on the outside left bank. In low-to-medium levels, this whole section calls for some fine maneuvering among holes and rocks. It should probably be scouted first. In high water (5.5), this stretch is Class 5, not due to the difficulty of any particular section, but because the current is very fast and unbroken. It is difficult to measure the gradient here, but it must be well over 100 feet/mile. With a swollen riverbed, this part of the South Branch is a swirling, boiling, cascading, stampeding beast, something best left to river gods and insane kayakers. Once past the second Route 12 bridge, there's a relatively straight-forward section until the river doubles under Route 12 again (in a left turn) and passes an island and a small roadside turnoff. In low-to-medium levels, these sections are Class 3; in high water they are Class 4, again because of the current's continuity. Along the way the paddler should watch out for small islands supporting tree growth: if you're not careful, you'll be wearing a new laurel wreath, roots and all. It's difficult to explain brush burns on your face when you've just come off a river.

Near the end of the trip, the South Branch makes a right turn and passes under Route 12 once more. The next left turn starts a "rock garden" which lasts for several hundred yards. It is fairly technical and about as far away from the road as the river gets. In high water the difficulty is only Class 4 because the gradient isn't so steep as it is upstream. One should not be lulled into a false sense of security, however, since there are plenty of exposed and unexposed rocks to jolt you back to reality. As these rapids wind down, you approach an old bridge and, very shortly afterward, a small dam, which you can run or not depending on water level. The dam is broken in the middle, although you can't tell this at high water levels. The gage is on the right bank here. This is also a convenient take-out spot, although boaters can easily continue down to and beyond the next left turn, which goes under Route 12 again. Beyond that bridge, the gradient gradually decreases, the river gets shallow, and, in low water, this section becomes a real drag. There is one spot downstream, however, where low-lying islands break the river into many narrow channels and the danger of tree hazards is high.

There is a U.S.G.S. gaging station on the right bank, just upstream from the dam at the take-out. There used to be a convenient outside staff, but it is no longer there. This is unfortunate because the river level readings for this description were correlated to it. There is now a hand-painted gage on the left, downstream side of the bridge just above the dam. The following level ratings are *estimates* based on the previous gage readings.

TOO LOW	LOW	MEDIUM	HIGH	TOO HIGH
0.5		1.5-2.0	2.5-3.0	

In the past, if the gage at Gilsum on the Ashuelot read 7.5, the South Branch was judged to be HIGH, if the Gilsum gage read 6.0, the South Branch was MEDIUM.

Bantam River (CT)

STODDARD ROAD TO SHEPAUG RIVER

Distance (miles)	Average Drop (feet/mile)	Maximum Drop (feet/mile)	Difficulty	Scenery
5.4	25	32	1-2	Good

TOO LOW	LOW	MEDIUM	HIGH	TOO HIGH	Gage Location	Shuttle (miles)
	0.5				Route 47 Bridge	5

The Bantam is traditionally one of the early trips. Along with the Shepaug, it nestles in the western Connecticut hills. Several groups usually run both rivers on the same weekend. The Shepaug is bigger and far more exciting. The Bantam provides an easy introduction to the Shepaug since it is Class 1-2 depending on the water level. To a hardy closed boater, the Bantam is as exciting as riding the crest when someone pulls the cork of your bathtub. All foolishness aside, however, the water should be respected, because in March it is very cold, and a swim could dampen your enthusiasm for later trips.

To start a trip on the Bantam, proceed south on State Route 25, from its junction with Route 209, for about 0.4 miles and turn left on the West Morris Road. Continue for a little less than a mile and take a left onto Stoddard Road, which leads to a bridge crossing the river. Here the Bantam is 2 to 3 boat

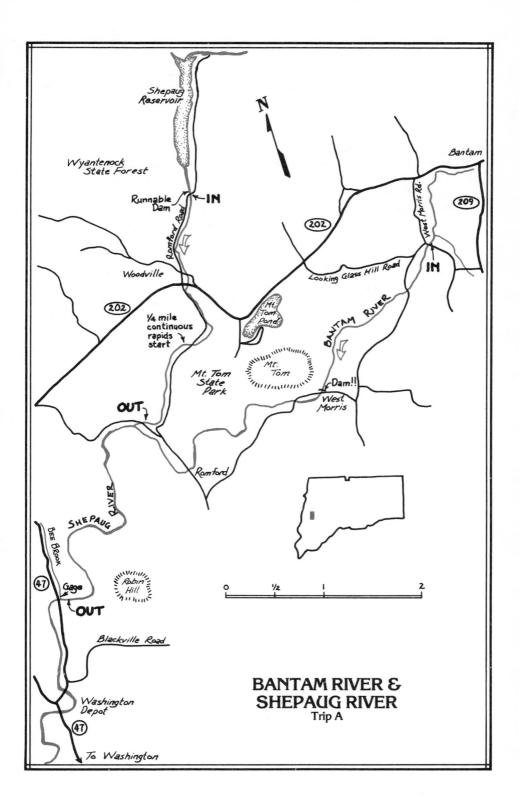

Shepaug
Reservoir

Wyantenock
State Forest

Runnable
Dam

IN

Romford Road

N

Bantam

West Morris Rd.

209

202

Looking Glass Hill Road

IN

Woodville

202

¼ mile
continuous
rapids
start

Mt.
Tom
Pond

BANTAM RIVER

Mt. Tom
State
Park

Mt.
Tom

Dam!!

West
Morris

OUT

Romford

SHEPAUG RIVER

Bee Brook

47

Gage

Robin
Hill

OUT

Blackville Road

Washington
Depot

47

To Washington

0   ½   1   2

BANTAM RIVER &
SHEPAUG RIVER
Trip A

lengths wide, with rapids above the bridge and smooth water flowing below. The river continues flowing smoothly most of its length, and it takes very high water to make it anything but an easy Class 2. In 0.8 miles, after a bridge, the river widens a bit but becomes no harder. The main danger in this stretch is downed trees, of which there can be many.

Just upstream from the bridge in West Morris, there is a broken dam which drops a total of 3 to 4 feet. There are many rocks at the bottom which may be covered at high levels and will definitely complicate an attempted run. It is probably best to carry. The river on the downstream side is slightly faster than above, and in high water there are even little ripples. Another set of standing wave rapids sits just in front of the next bridge and there's also one after the bridge.

The Bantam then stays smooth. One right turn has some little rapids and below, in a left turn, there's a six-inch ledge. Downstream a short distance is a right turn with a large rock in the left center. Below that there's another drop over a small ledge. These ledges are gone in high water. The next landmark is the steel bridge at Romford, which is about 1 mile from the confluence with the Shepaug. Once the Shepaug is entered, the tempo picks up. For details, see the Shepaug description.

Bearcamp River (NH)

BENNETT CORNER TO WHITTIER

Distance (miles)	Average Drop (feet/mile)	Maximum Drop (feet/mile)	Difficulty	Scenery
3.5	32		2-3-4	Good

TOO LOW	LOW	MEDIUM	HIGH	TOO HIGH	Gage Location	Shuttle (miles)
0.5	1.0				Whittier	3.5

The Bearcamp is an interesting river located just south of the White Mountains. Depending on water level, it can be run by open or closed boats, although open boaters will have to exercise caution in several spots. The major rapids are well defined, each is less than 100 yards in length, and they are separated by stretches of flat or Class 1-2 water. There is one dam that requires a portage, and two rapids may need scouting. Route 25 parallels the river, but is not obvious enough to detract greatly from the scenery.

Start the trip west of South Tamworth by turning off Route 25 on to Route 113, where a bridge crosses the river. There is also a Route 113 turnoff in Whittier. At the put-in the Bearcamp is about 60 feet wide with a Class 1 current and a sandy bottom. Proceeding downstream, the riverbed meanders gently with trees possibly blocking part of the passage; since the current is mild, they should present little hazard. After about 15 minutes' paddle, the first rapids is seen; it is Class 2+ at low water, Class 3 at higher levels. The total length is around 100 yards, and there are an abundance of small rocks to maneuver around. There are many routes through to the pool that sits at the bottom. A rock ledge across the river follows very shortly. The left side is easier; the right side has a more abrupt drop. Class 2 rapids interspaced with flat water follow. Soon, Route 25 approaches the right bank, and several houses come into view.

The quiet-flowing Bearcamp terminates in Cold Brook Rapids (Class 3-4). Starting some 100 yards above an iron bridge, the current speeds up and large rocks line the sides. The water is choppy and halfway down to the bridge a large boulder in center stream forces most of the current to the right where the heaviest water is located. If you stay with the main flow, Cold Brook Rapids is fairly straightforward, although there are many little drops leading to the bridge and some nice haystacks underneath. Just above the bridge, on the right, Cold Brook picturesquely cuts its way through solid rock to join the Bearcamp. There is a big eddy on the left, downstream side of the bridge which you should pop into before attempting the final part. The rapids continue on past the bridge and, in about 30 yards, there's a sudden drop over a ledge. A rock in the center of the ledge defines two channels — right or left. The left side drops a bit more sharply (3 feet) than the right, but the right channel has several small rocks at the bottom, just waiting to push your bow into the cockpit. At a gage reading of 1.0, this whole stretch is Class 3, harder at higher levels. Scout it if in doubt. Open boaters shouldn't feel bad about walking around. The outflow travels down to, and around, the next turn, where the Bearcamp Gorge can be seen.

The Bearcamp Gorge is enshrouded by vertical rock walls rising 15 to 20 feet. It lasts 50-75 yards and is S-shaped. The entrance is via a sharp right turn where the current rushes hard against the outer rock wall. In the left turn that completes the S, there's a 2- to 3-foot abrupt drop (low water) steepest on the outside right. This whole section is narrow, fast, and turbulent (Class 3+ at LOW water). A large pool below is handy for picking up the pieces. Open boaters should treat these rapids with caution. It is best to run solo. Scout before running, as you could be up the creek if a tree is down.

Below the pool is a broken dam that should be portaged, although "Krazy Kayaker" may want to run the narrow 5-foot drop on the extreme right. Immediately below the dam an island divides the river. Flat water and Class 2 rapids follow.

The next goodies, just upstream from a small bridge, are a series of small ledges. Route 25 is alongside on the right, and the river is shallower than above. The last eventful rapids is just before the Route 113 bridge in Whittier. It consists of a narrow S-shaped channel with large rocks on the sides and smaller ones in the water. It's Class 3 in low water, and could be exciting in medium or high water.

There is a hand-painted gage on the left, upstream side of the Route 113 bridge in Whittier.

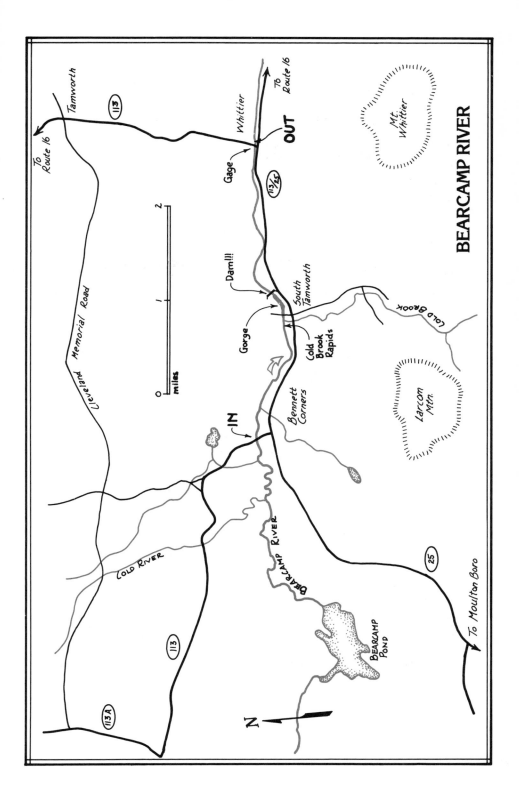

BEARCAMP RIVER

Mt Whittier

Larcom Mtn.

To Route 16

Tamworth

113

To Route 16

Cleveland Memorial Road

2

1

miles

0

IN

Bennett Corners

Cold River

113

113A

Cold River

BEARCAMP RIVER

BEARCAMP POND

25

To Moulton Boro

N

Gorge

Dam!!

South Tamworth

Cold Brook Rapids

Cold Brook

113½

Gage

Whittier

OUT

To Route 16

Black River
(VT)

WHITESVILLE TO PERKINSVILLE

Distance (miles)	Average Drop (feet/mile)	Maximum Drop (feet/mile)	Difficulty	Scenery
7.5	27	40	2+	Fair

TOO LOW	LOW	MEDIUM	HIGH	TOO HIGH	Gage Location	Shuttle (miles)
0.5	1.5				Covered Bridge	7.5

The Black is a medium-sized river in southeastern Vermont which offers a straightforward trip for advanced beginners. The usual run depends to a large extent on how much water is being released at the power station in Cavendish. A road parallels the river, so much of the trip can be viewed prior to launching. The river valley is not particularly attractive — even less so during hunting season with abundant red-coated bipeds running through the woods. The riverbed is larger than most, so there's much room for what little maneuvering is needed. The rocks are mostly small, although several sections do have boulders that would be rather tough to displace with a moving canoe. In high water, the current can be fast, although that would be the only danger. The most distinctive aspect of the Black, however, is the *unrunnable* Cavendish Gorge.

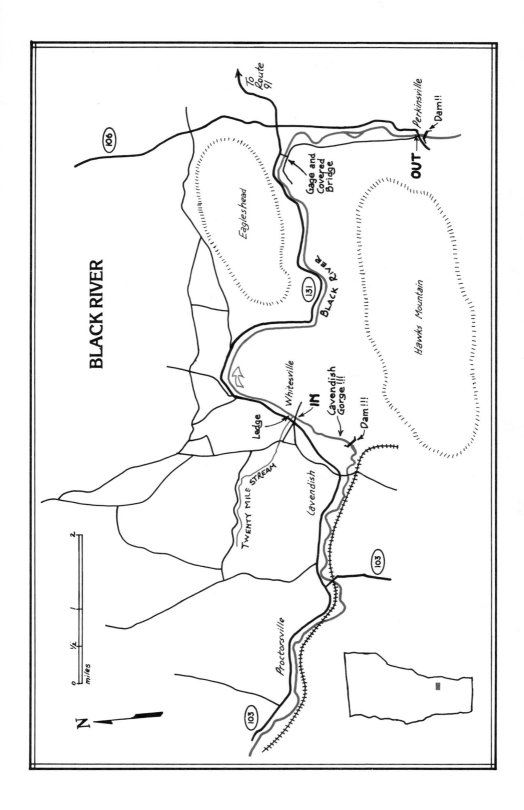

Blocked on its upstream end by a 30-foot dam, the Cavendish Gorge extends some ¼ to ½ mile to the power station at its base. If this place is an example of how Mother Nature cuts through things, you definitely don't want to let her carve your Christmas turkey. Vertical rock walls spaced 20 to 30 feet apart and up to 50 feet high, abrupt 90-degree turns, undercut passageways and ledges, and huge semi-spherical pot holes gouged out of solid rock typify this stretch of Class 7 water. You would have to be mad to attempt this in a boat: there are channels through which a boat couldn't even fit.

After visiting the Cavendish Gorge, launch your boats at Whitesville, which consists of several houses that have seen better days. Turn off Route 131 onto Carlton Road (by a Cavendish Mills sign) and drive 50 yards to the Black where a bridge crosses the river. The Black is fairly narrow here, and it flows with a good current.

Immediately downstream from the bridge is a ledge that extends all the way across the river. This is the most difficult single rapids on the trip. In normal water, there are passages at either extreme; the left channel has a less abrupt drop. From the bridge, approach the ledge in the right or right center and, when the current splits, stay left. The total drop is around 2 feet. The left side is narrow, and almost anything can block it, so look closely before you run. The drop in the right channel is abrupt and requires a sharp turn to enter. Higher points of the ledge block the middle of the river. A pool follows shortly and Twenty Mile Stream enters from the left. If you don't like the looks of this ledge for starters, put in farther downstream, where Route 131 come close to the river.

For the next mile the Black is Class 1-2 in difficulty. You pass an old section of Route 131; then there's another mile of Class 2 water.

About 3.5 miles into the trip, the paddler should be on the lookout for rocks that are larger than normal. They mark the beginning of a section that is a bit more difficult. At one point in this section very large rocks line the banks, narrowing the riverbed and creating a fast Class 3 turbulent channel where maneuvering is necessary.

Approaching a covered bridge, small rocks force some easy turning in low water. After this bridge, the Black turns right and less interesting paddling commences. A take-out by the covered bridge is probably best. Just below the bridge in Perkinsville, in a slight right hand turn, there is a dam.

There is a hand-painted gage on the left, upstream side of the covered bridge near the take-out.

Blackledge River (CT)

ROUTE 66 TO SALMON RIVER

Distance (miles)	Average Drop (feet/mile)	Maximum Drop (feet/mile)	Difficulty	Scenery
5.9	20	30	2	Good

TOO LOW	LOW	MEDIUM	HIGH	TOO HIGH	Gage Location	Shuttle (miles)
1.4		3.0			Comstock Bridge	5

    The Blackledge is a narrow, winding stream that is sometimes used for entering the Salmon. Since it has a very small watershed, it is up only during heavy spring runoffs. When it's up, standing waves are the order of the day. When it's low, rocks appear, making a nuisance of themselves. At this lower level the Blackledge is not worth the effort.

    From the junction of old Route 2 and Route 66 in Marborough, Connecticut, proceed about 1.5 miles east on Route 66 to a spot where a small bridge crosses the river. Here the river is about two boat lengths wide, and the current's speed depends on the water level. Put in on the upstream side of the bridge, as the lower part has been posted against trespassers. An alternative starting point is along old Route 2 where the river approaches the road. Here, there is a small dam crossing the stream and several closely spaced meanders. The entire river is narrow with no special difficulties other than tree hazards and the narrowness, which restricts maneuvering. Approach curves with caution, because a tree could be down almost anywhere. From Route 66, the Blackledge runs for about 3.3 miles through the Salmon River State Forest, away from roads, before it approaches both old and new Route 2. After it finally leaves Route 2, it again runs away from roads for 1.5 miles, where it joins with the Jeremy to form the Salmon.

The gage for this trip is on the Salmon at Old Comstock Bridge. For further details, see the Salmon description.

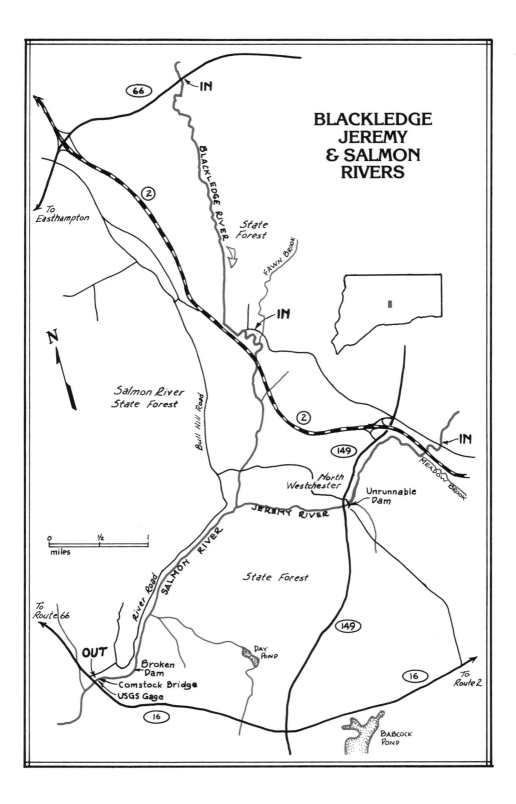

BLACKLEDGE
JEREMY
& SALMON
RIVERS

66

IN

BLACKLEDGE RIVER

2

To
Easthampton

State
Forest

FAWN BROOK

IN

N

Salmon River
State Forest

Bull Hill Road

2

149

North
Westchester

IN

MEADOW BROOK

Unrunnable
Dam

JEREMY RIVER

0   ½   1
miles

River Road

SALMON RIVER

State Forest

149

To
Route 66

DAY
POND

OUT

16

To
Route 2

Broken
Dam

Comstock Bridge

USGS Gage

16

BABCOCK
POND

Blackwater River (NH)

ROUTE 127 TO SNYDER'S MILL

Distance (miles)	Average Drop (feet/mile)	Maximum Drop (feet/mile)	Difficulty	Scenery
2.5	24	50	1-4	Fair

TOO LOW	LOW	MEDIUM	HIGH	TOO HIGH	Gage Location	Shuttle (miles)
	3.4C	4.3C			Webster	2.5

The Blackwater is a small, little-known stream that offers a short, racy run between slalom gates made of summer homes. The river is dam-controlled, so it is possible to know exactly how much water to expect on any one weekend.* Looking at the gradient figures, one would suspect the Blackwater cannot offer much white water, and, in places, it doesn't. There is the flat water, and there are the rapids, and the two are quite distinct. In the upper half of the trip, the rapids are well defined and short. In the lower half, the rapids are narrow, technical, fast, and they tend to blend together, forming several continuous thrillers. The Blackwater is not far from Concord, New Hampshire, and can serve as a follow-up run to the Contoocook.

*The AMC gets weekly river readings from the U.S. Army Corps of Engineers. See page 27.

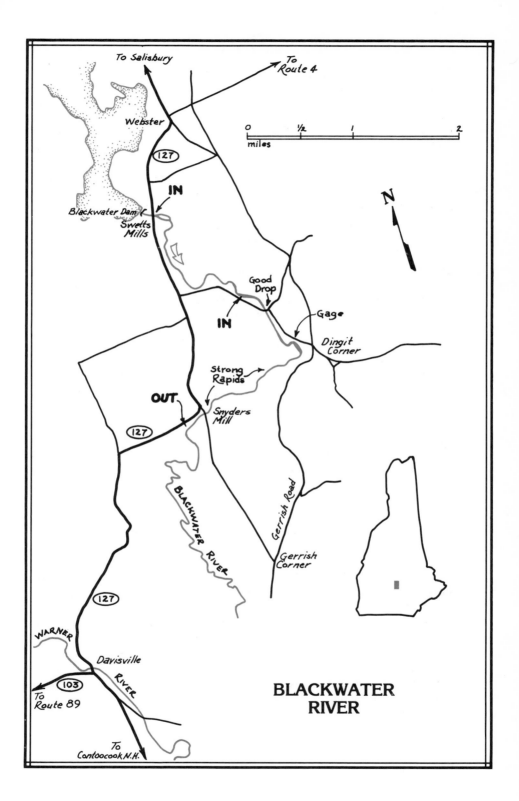

To Salisbury

To
Route 4

0   ½   1   2
miles

Webster

127

IN

N

Blackwater Dam
Swetts
Mills

Good
Drop

Gage

IN

Dingit
Corner

Strong
Rapids

OUT

Snyders
Mill

BLACKWATER RIVER

Gerrish Road

Gerrish
Corner

127

WARNER

127

Davisville

RIVER

103

To
Route 89

To
Contoocook, N.H.

**BLACKWATER
RIVER**

Start a trip from a small side road that connects with Route 127 and approaches the river. At the start the river is relatively calm, and parts of it are pool-like in low water. If this spot proves difficult for a put-in because of posted land, an alternative start is on the right, downstream side of the Route 127 bridge. The bridge is just downstream from Blackwater Dam. The outflow from the dam is rather dramatic, and, below the bridge, the river forks around an island in Class 2-3 fashion.

When Clothespin Bridge (the name is bigger than the structure) is spotted, the boater should prepare for a real neck-jerker. Just before the bridge, the riverbed narrows and drops precipitously about 4 to 6 feet into an even narrower channel which can cause real problems if it is not run just right. In the approach, the boater first paddles some calm water; then comes a short stretch of a slightly S-shaped Class 2-3 rapids that terminates in the big plunge. At the bottom, a series of boulders comes out from the left bank forcing some furious last-second maneuvering in mid-air. It is very difficult to see the details of the drop from boat level so your memory and instincts must be your guide. People have run this drop at various levels, but each boater should scout and judge for himself. The whole scene can be viewed from the bridge, but be aware that the surrounding area is heavily posted against swimmers, waders, parkers, loiterers, procrastinators, and urinators. The outflow from the drop down to the bridge is fast and tight. Below the bridge there's a little rocky rapid that splits around a tiny island. Calm water follows.

From Clothespin Bridge onward, there are many summer homes, some with angry dogs in residence. In less than a half mile of easy paddling, the gage is visible on the left bank. The gage has a dual purpose: it reads the river level, and it marks the beginning of the lower half of the run. From here on, there is either flat water or good white water. Both are obvious when spotted. Almost directly below the gage the river turns right and falls over a short rock pile which is very scratchy at a reading of 3.4. Beyond, the water turns flat.

The Eggbeater is the second of two long sets of rapids in the lower part of the trip. The first set, called "A" rapids, is about 100 yards long. If you think "A" is hard, wait until you encounter Eggbeater. There is some flat water between them. Eggbeater is easily recognized because the whole river suddenly turns white like a bowl of meringue beaten to a froth. It starts in a right hand turn, and soon narrows into an extremely fast, tight, rocky chute that has the boater aiming first at one bank, then at the other. Turbulence, holes, and an extremely powerful current combine to make the Eggbeater a Class 4, even in low water. On occasion, even good boaters manage to get turned around and are forced to run backwards. This is certainly not recommended although it does have an advantage in that you can't see what's coming next. Any swim you take here will probably be a long, brutal one. At the end of it you qualify for the Humpty Dumpty award. Eggbeater terminates rather dramatically in a right turn, just upstream from Snyder's Mill Bridge. A series of ledges crosses the river in this turn. Depending on where you cross their line, you'll have an easy or a hard time of it. The steepest drops tend to be on the inside part of the curve. Avoid the right side if you can. The ledges continue beyond the bridge, with the rapids growing easier all the while.

Take out somewhat downstream of Snyder's Mill Bridge on the right bank. A small sandy road leads to Route 127 where there is a house (a cape with a bay window) and a small store.

The gage is on the left bank, 0.2 miles west of Dingit Corner. It is hard to find from the road. Make that impossible to find.

Boreas River
(NY)

ROUTE 28N TO MINERVA BRIDGE

Distance (miles)	Average Drop (feet/mile)	Maximum Drop (feet/mile)	Difficulty	Scenery
7.0	45	100	4-5	Excellent

TOO LOW	LOW	MEDIUM	HIGH	TOO HIGH	Gage Location	Shuttle (miles)
4.8C		5.5C	7.0C	7.0C	North Creek	11

There are only a few rivers whose reputations do not overshadow actual fact. Most paddlers commonly exaggerate the difficulty of rivers and rapids to the point that the unsuspecting listener really doesn't know what to expect. There are, however, a few waterways in which the horror stories, the tall tales, and the descriptions are not quite so padded with campfire chatter. These are a special breed of classic rivers which live up to the tough reputations they have. The Boreas is one of these.

Born in Boreas Pond somewhat northeast of Newcomb, the Boreas flows through Cheney Pond and then generally southward to rendezvous with the Hudson. For most of its length, the watershed valley is extremely attractive, resembling paintings from the Hudson River School. Towering conifers line the banks, and since the valley is mildly inaccessible, the scenery is generally

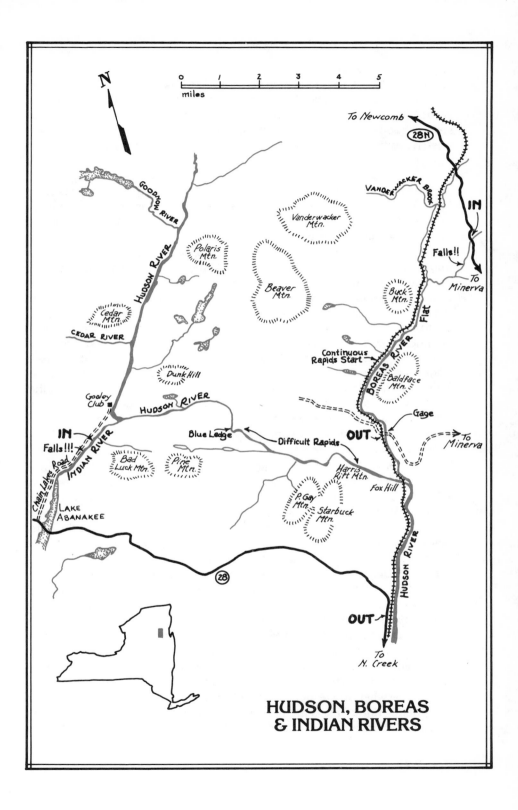

HUDSON, BOREAS
& INDIAN RIVERS

unpierced by civilization. Through the middle part of the run, the Boreas is absolutely flat, reflecting the blue sky and white powder puff clouds, as if they were somehow originating beneath the watery surface itself. The rapids are something else again. Two sections are noteworthy. The upper part has several good sets of rapids, terminating in a runnable waterfall, and the lower stretch has the longest, most continuous length of difficult rapids discussed in this guide.

The usual starting point for a trip, or what is more correct, a happening, on the Boreas is the Route 28N bridge which crosses the river northwest of Minerva. Here the Boreas is some 50 feet wide, and although it is flowing at a fair clip it presents a rather mild-mannered facade. To the east of the bridge a dirt road pushes its way upstream into the woods, leading to a campsite area and a convenient launching spot. An island splits the stream here giving the impression that one could spit across the whole river. Downstream, beyond the bridge, the Boreas turns right where a few rocks and ripples appear. The quiet water continues long enough for the paddler to limber up muscles and to get comfortable in a favorite position. The first real rapids begin in a very sharp left turn where at the start rocks leapfrog out from the left bank, narrowing the channel. This stretch lasts about 100 yards. It is Z-shaped and tight, with numerous small drops throughout. This natural slalom's lower part is a sharp right turn where the main current is on the outside against some rocks. Immediately following is a short (15 yards), narrow (15-20 feet) chute that terminates in another abrupt drop with a smoother tongue located on the extreme right side (low water only). If this rapid seems tough, the paddler may be well advised to shore the canoe and carry it back to Route 28N, since those below are much more severe.

Shortly downstream are several more small drops over rock ledges and then there's a waterfall. Dropping precipitously on either extreme (8 feet), the falls have a smoother, two-stage descent in the middle; the first stage is more abrupt than the second. These falls have been run in both high and low water, but boaters should judge for themselves. The portage is easiest on the right. Below the falls, in a sharp left turn, there is another chute with a stern-kicker at the bottom. At low levels, the outflow is rocky but easy. An island forces the current 90 degrees to the right or left, after which there is smoother water.

Class 2 water continues until the railroad makes its initial approach on the right shore. Here Vanderwhacker Brook enters from the right, and the Boreas turns 90 degrees left. For about the next three miles, the course is smooth and mostly flat, past several islands and some small overhead cables. In this section the Boreas is at its widest, being 50-75 feet. The sleepy peacefulness and quiet tranquility of the scenery will impress even the most ardent city dweller. However, as always, there is calm before the storm.

Just when you're tired of flat paddling, notice a faster current as the river turns gently right. This is the only warning before all hell breaks loose. What follows is 2.5 miles of violent and absolutely uninterrupted difficult rapids, too complex to describe in detail. Beyond that right turn, the banks crowd together creating a natural sluiceway, and the whole riverbed tilts downward as if trying to throw the paddler off its back. This 2.5 mile stretch averages about 90 feet/mile gradient and achieves a maximum of over 100 feet/mile. At first, rocks are just numerous, but they soon line every conceivable space as the current stampedes uncontrolled over and around them, causing countless haystacks, hydraulics, and dizzying crosscurrents. Needless to say, the turbulence is something more than placid. This mayhem continues for a long time — at least the paddler will think so. In one spot, where the railroad is close by on the right, a series of rocks thrusts out from the left bank, narrowing the already narrow channel to that of a boat's length. Just upstream, and lasting perhaps 5 to 10 yards, is a small, quiet pool. On its left is one of the few usable eddies to be found on the trip. In high water forget about the pool — it's gone. If you can get into the eddy, however, it's a good place to rest and survey what's below.

What follows is more of the same, only harder — another long section of continuous, extremely difficult technical maneuverings. In this section, as in the one above, the current can change direction without notice, forcing the canoeist to make instantaneous decisions time after time. Downstream, there are two boulders, one in the left and one in the right center. Between them is an abrupt 3-foot drop. Immediately below, the Boreas turns sharply left and forms a small eddy on the outside bank among some rocks. This introduces the famous Z-turns. Superimposed on the other difficulties are several *very* sharp turns, each choked with rocks. If you get by these goodies without broaching or bouncing down on your head, consider yourself a respectable boater. The river continues to dish out abuse, although it diminishes somewhat just above the Minerva bridge. Along the way you'll pass an island where there's a strong hydraulic on the left side. Once you spot the bridge, the hardest is over. If you choose to continue to the Hudson (2 miles from the bridge), the trip will gradually wind down. After 3.5 miles on the Hudson, Route 28 appears on the right.

Alternatively, cars can park by the Minerva bridge, reached by driving 1.8 miles north on Route 28N past the general store in Minerva, then taking a dirt road that slants off to the left, just past a rustic lodge on the right. After only four miles on this Class 2-3 (Class 4 if it's wet) road, the Boreas bridge is reached. In spring this road will most probably be blocked by snow or mud.

There are two gages that indicate the canoeability of the Boreas. The easiest to read, since it is in the Telemark system, is the one in North Creek on the Hudson. Unfortunately, this gage is on the Hudson, not the Boreas, and the correlation is only a very general one — if the Hudson is up, the Boreas is likely to be up also. The other gage is on the Minerva bridge; it is hard to read because of its location. If it reads 1.0, you should have an exciting run.

Some additonal comments should be made concerning this run. Once started, the last rapids must be completed one way or another. More than one unlucky paddler has used the railroad that follows the river to walk out after losing a boat. If by chance you are forced to leave your boat, the swim could be awesome: it could be your last if the water is high. In high water people have reported being swept downstream for what they estimated to be a half mile or more before getting out. High levels are rated at least Class 5, and, although the rocks are mostly covered, the pace is amazingly fast and turbulent. In low or medium water, the Boreas is rated Class 4. The Boreas is no place for an open boat at any level. For those who can handle it, this run is excellent sport, but it should not be taken lightly.

Branch - Pascoag Rivers (RI)

HARRISVILLE TO GLENDALE

Distance (miles)	Average Drop (feet/mile)	Maximum Drop (feet/mile)	Difficulty	Scenery
4.8	8	40	1-2	Good

TOO LOW	LOW	MEDIUM	HIGH	TOO HIGH	Gage Location	Shuttle (miles)
2.9					Forestdale	3.5

Although Rhode Island is not known as a white water haven, it would be erroneous to assume that there is no river running to be had in this state. What the Pascoag-Branch Rivers lack in white water thrills, they make up in idyllic beauty and serenity. Flowing along low banks, thickly settled with trees, these small rivers meander endlessly, as if they were trying to tie themselves into knots. The current is usually present, though it is weak in spots. Rapids are easy riffles and several abrupt drops over small debris-laden dams. There is also one section of several hundred yards in Glendale where the pace increases and the boater faces some rocky Class 2 rapids. Since the rivers are so narrow, trees can completely block the path, so trees are a hazard throughout the trip. These rivers are definitely for beginners or for those seeking an amiable commune with nature.

Start the trip on the Pascoag River, just downstream from the dam in Harrisville. Immediately upstream from the Route 107 bridge is a collection of rocks and stones that can be negotiated or not depending upon the water level. If the gage is 2.9 or lower, carry your boat under the bridge and put in there. After passing a factory on the right bank, the water sets the pace — slow and even. After several turns, there is a ½- to 1-foot drop over a small dam that could easily change complexion every year. There are now woods on the right bank and houses on the left. After you pass the remains of an old bridge, woods populate both banks.

Following a right turn, there waits an old dam with a two-stage descent. In low water it is scratchy to get over; in medium water the drops should be easier. Look it over if you are unsure about where to go. There are some big boulders below, but they are easy to miss.

The Pascoag continues on in its typical sinuous manner, under two bridges and past the remains of several more. When the river turns sharply left and passes some old railroad bridge supports, the paddler enters a marshy area where the channel is deeper and wider than normal. This is where the actual Branch River is formed by the confluence of the Pascoag and Chepachet Rivers. Also, this is the beginning of some backwater behind a dam. This dam is near a factory on the left bank; the portage should be made on the right bank. Below the dam, the current is quicker than normal, but after the first turn the river reverts back to its characteristic temperament.

As you approach Glendale, a row of old bridge supports spans the river. You'll have to pick and choose your way through the debris piled up on them. Next are two closely spaced bridges; then the boat enters a section of continuous fast current that is peppered with rocks. This portion of the river starts by dropping over a small broken dam, passes to the left of an island, then continues down past more rocks and waves. This is probably the most continuously difficult part of the trip; at a gage reading of 2.9, it is a very scratchy and annoying Class 2 in difficulty. This stronger current lasts for several hundred yards.

Several take-outs are possible. The first is in Glendale, just at the end of the set of rapids there. A factory (Bruin Plastics) and also the Glendale fire and rescue station are in view on the left bank near this spot. An alternative is to continue to the junction of Routes 7 and 102 and take out there. If you do this, be careful of the dam just upstream from the Route 102 bridge.

The gage is in Forestdale, about 400 feet downstream from a mill dam. The gage is at the bottom of a steep hill behind the H & H Machine Tool Company, which is itself down and across the street from the Forestdale Post Office. The 2.9 gage reading is rated LOW for most of the trip where there are no rapids, and it is rated TOO LOW for the section of rapids in Glendale. At this level, passage is possible, but it will lacerate the bottom of your boat.

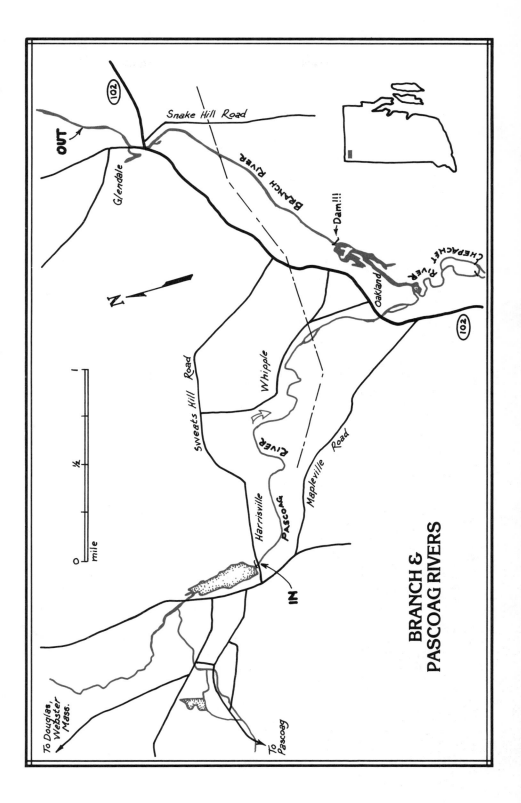

BRANCH &
PASCOAG RIVERS

Chickley River (MA)

ROUTE 8A TO DEERFIELD RIVER

Distance (miles)	Average Drop (feet/mile)	Maximum Drop (feet/mile)	Difficulty	Scenery
4.4	77	100	2-3	Good

TOO LOW	LOW	MEDIUM	HIGH	TOO HIGH	Gage Location	Shuttle (miles)
4.3		5.3C	5.3S	5.3	Shattuckville	4.0

The Chickley is a little-known and little-canoed stream that runs into the Deerfield near the mouth of the Cold River just west of Charlemont. Although relatively small at the start, it broadens as it progresses. It can pack a big punch with almost nonstop rapids and several good drops in the upper part and somewhat more relaxed rapids and another good drop in the lower part. In high water, standing waves measuring 3 feet in height are not uncommon. It is at this level that the Chickley presents the biggest challenge and excitement. At lower levels, the Chickley will be a typical small river, Class 2-3.

Put in at a bridge above the Chickley Alps Ski Area, about 4 miles from the mouth. Looking above the bridge, the river comes roaring down the mountainside only to tumble into Upper Rapids just below the bridge. Upper Rapids consists of two parts, each with a rather abrupt drop. In the lower part,

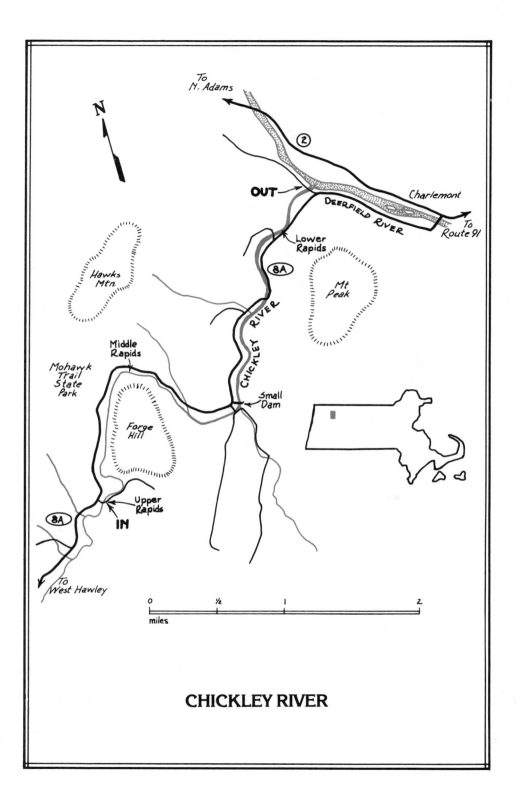

CHICKLEY RIVER

water falls over a 1- to 1½-foot dam, then eventually turns right. If you start out at the bridge itself, you will enter these rapids quickly. An alternative is to put in from the right bank just above the dam where there is a small quiet spot. Once past Upper Rapids, the next quarter mile downstream is almost continuous rapids in high water, with several places that are quite interesting. In high water, most rapids are standing waves but the strong crosscurrents and angled hydraulics that seem to exist everywhere make this run very eventful and tricky. At lower levels, this whole section would be great practice in tight rock dodging. At one point look out for a large boulder on the right side; beyond it are two long, heavy sets of rapids separated by a bit of calm. Next, there are some good standing-wave rapids where the Chickley constricts a little in a right turn. Since the river is narrow, always be careful of downed trees.

When you see the road approaching on the left and the river turns right below, get ready for the next test—Middle Rapids. Three rocks just in the turn, two small ones on the right and one bigger rock in the center, guard the approach to this innocuous looking but very fast rapids. In high water, pass just to the left of the big rock, if you can, and stay to the extreme left of the main channel to avoid an ugly (and also unseen from above) souse hole at the bottom center and right. A center route is certainly possible, but you would hit the souse hole directly. These rapids are some 25-50 yards long and can be seen from the road — it's a good idea to scout Middle Rapids. On the north side of the road 500 yards or so downstream is a farmhouse that you can use as a scouting landmark, though it's not visible from the river. This rapids is approximately 1.2 miles from the start. In lower water, pass to the right of the center rock and proceed down the right center or right, avoiding dragon teeth as they appear. Move left halfway down, then continue straight. At this low level, the left side is an unpassable mass of rocks. A tree across any part of the river here would put the paddler in real trouble. Continue on with quick water and standing waves to the second bridge, below which on the right a small stream enters. Below is a small rapids and another 1- to 2-foot dam that can be run straightforwardly. After this, the river's pace noticeably slackens. In high water a good Class 2 trip could be started from this second bridge.

A word of caution should be given about fences across this river. On the scouting trip, several were encountered, one of which appeared to be electric from its construction. One fence was just below the third bridge, and another was 200 to 300 yards below this same bridge, where a small island divides the channel. On future trips any or all of these may have disappeared, but be on the alert against the possibility. Coming upon one of these unexpectedly can cause a certain amount of apprehension to most paddlers. From a boater's standpoint, wire fences across the river are extremely hard to spot, and they are very dangerous.

A short distance below the fourth bridge is the last heavy rapids—Lower Rapids (Class 3-4 in high water). Pull out well above if you're going to scout, since the current leading into it is very fast. The rapids itself is over in 15 to 20 feet and the canoeist may well be also, because several crosscurrents and heavy hydraulics compete for control of the canoe. In lower water, it is evident that Lower Rapids has a double ledge. The first drops 1 to 2 feet, while the second drops .5 to 1 foot, about 10 feet downstream from the first ledge. The main current flowing out of the second ledge runs into an undercut rock on the right side, so think left side after passing over the first ledge. The Chickley is only one boat length wide at the bottom of the second ledge, so don't try this one sideways. From here to the take-out, the Chickley is a relatively easy Class 2, but still has a strong current.

The Chickley has no known gage. On the scouting trip, the water was high due to an unusual thaw in the beginning of January. That day, the North River's gage at Shattuckville read 5.3. It was estimated that the Chickley was running about 500-600 CFS at its confluence with the Deerfield. At this level, open boats would take in a great deal of water running the rapids on the upper stretch, so this is not recommended. On another trip, the North's gage read 4.3, and the Chickley's canoeability was rated between TOO LOW and LOW. At the lower levels, the upper section is good sport for all in the Class 2-3 range, with countless rocks, ledges, and hydraulics. The HIGH water level is rated a hard Class 3 or an easy Class 4. A road parallels the river most of the way, near at hand should you need it. The Chickley serves well as a replacement for the rather bottled-up Deerfield, and it is definitely more sporty.

To reach the mouth of the Chickley, pass west through Charlemont, turn onto Route 8A, cross the Deerfield, then turn right. Continue for about 2.2 miles until the road splits. Take the right fork and in 100 yards you will find the fifth bridge crossing the river. To launch, take the left fork (West Hawley Road).

Cold River
(NH)

SOUTH ACWORTH TO VILAS POOL
TRIP A

Distance (miles)	Average Drop (feet/mile)	Maximum Drop (feet/mile)	Difficulty	Scenery
5.5	43	53	2	Good

TOO LOW	LOW	MEDIUM	HIGH	TOO HIGH	Gage Location	Shuttle (miles)
3.4		5.2	8.0		Drewsville	5.5

The Cold is a small, snappy stream which offers nearly continuous current, pleasant scenery, fallen trees, and barbed wire. In low water the pace is relatively slow, while in medium to high levels the Cold offers a nonstop run. The Cold is an excellent place to practice soloing in an open canoe, because while none of the rapids are extremely difficult, they do require a certain finesse to negotiate gracefully. The valley is alternatively attractive and civilized, and the water is polluted only by furry beasties and cows. When canoeable, the water temperature lives up to its name. A road passes alongside almost the entire route, so starts and stops are flexible.

Put in below the gorge, downstream from South Acworth where the road comes close to the river. This gorge has several sharp drops, the last being a 10-foot waterfall. At the put-in, the river is 20 to 25 feet wide, moving

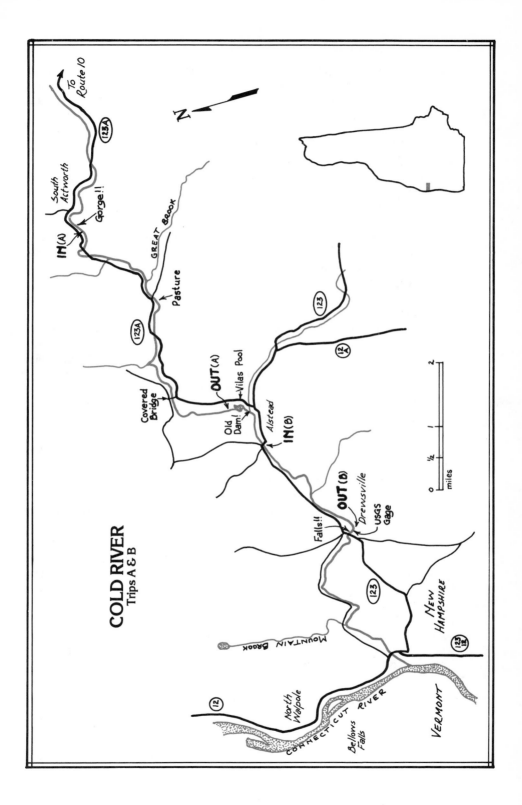

COLD RIVER
Trips A & B

To Route 10

123A

South Acworth

Gorge!!

IN (A)

GREAT BROOK

Pasture

123A

Covered Bridge

OUT (A)

Vilas Pool

Old Dam!

Alstead

IN (B)

123

12A

Drewsville

OUT (B)

USGS Gage

Falls!!

123

MOUNTAIN BROOK

NEW HAMPSHIRE

123 12

North Walpole

12

CONNECTICUT RIVER

Bellows Falls

VERMONT

N

0   ½   1   2
miles

swiftly over small rocks with a scattering of larger boulders. A mile or so downstream, the river turns left against a 5-foot rock on the right, and a bit downstream, a 6-inch rock dam should give no trouble. Further down, the stream narrows and seems to double back on itself in several sharp turns, where the risk of fallen trees is high. This section also suffers from obstacles, such as footbridges and makeshift dams. The boater should be particularly watchful for barbed wire, especially in the area of Great Brook, a side stream which enters from the left. A little bridge crosses Great Brook just before its confluence with the Cold.

Beyond the next bridge, the Cold flows through a pasture, where a curious cow or two (or twenty) may follow the funny-looking water animals. Passage to the following bridge has some crisp rapids and one left turn that is sharper than usual; otherwise there is nothing difficult. Under the Route 123A bridge, the river forks with channels on the extremes of both sides. Continuing downstream, the Cold offers more brisk rapids where the narrow riverbed forces the canoeist to plan ahead carefully, in order to zig and zag around rocks and fallen trees.

Near a covered bridge, again be on the lookout for more barbed wire fences. You may not find them, but consider that a good omen. About .25 to .5 miles below the covered bridge, the road leaves the river. At that point there is a series of particularly good Class 2-3 rapids, spaced between swiftly moving water. The first one, in a left turn, is one of the best sections of the trip. The danger of fallen trees in every turn is great. The pace accelerates and the river twists like a snake until the entrance to Vilas Pool, about 1.5 miles from the covered bridge. When you enter Vilas Pool, take out immediately on the left side where the road is near in order to avoid the dam on the downstream side of the pool. The outflow orifice of this dam is usually below the waterline, and it is just big enough to trap a body or a boat.

The Cold is runnable at almost any level, assuming, of course, that there is some water. Even at a gage reading of 8.0, the Cold is manageable in an open boat, but the current is strong and rescues will be difficult. At this high level, the rapids are almost all standing waves with few or no rocks to be seen. The sideline trees are also in the water, which makes eddying out more difficult. At a reading of 8.0, the Cold is rated a heavy water Class 2 or an easy Class 3. The route is usually straightforward but Class 2 boaters who are just starting out should exercise caution at these higher levels.

The gage is located in Cheshire County, on the left bank, 50 feet upstream from the bridge on State Highway 123A, north of Drewsville. The gage is just upstream from a short but impressive gorge where the Cold plunges down an almost vertical staircase.

Cold River
(NH)

ALSTED TO DREWSVILLE
Trip B

Distance (miles)	Average Drop (feet/mile)	Maximum Drop (feet/mile)	Difficulty	Scenery
2.0	40	40	2	Good

TOO LOW	LOW	MEDIUM	HIGH	TOO HIGH	Gage Location	Shuttle (miles)
3.4		5.2	8.0		Drewsville	2.0

This trip may be made in conjunction with the one above or as a separate trip. Just below the dam on Vilas Pool, immediately after a right turn, there is another dam. This dam is broken and may be runnable depending on the water level and your ability. Look it over and decide for yourself. Since you have to portage the dam at Vilas Pool, you might as well carry down to Alsted if you want to continue on from the upper part (Trip A). Because Trip B is only 2 miles long, it presents a good opportunity for new canoeists to practice, without having to tackle the longer upper section of the river.

This section of the river is similar to the preceding section, and, although the rapids diminish somewhat, the swift current continues. The river is still narrow and fallen trees can aggravate the paddler and slow the trip drastically. Approach each blind curve cautiously, looking for obstacles.

Beyond the trees, there are no major difficulties—except for the take-out.

The trip ends as the river takes an awesome plunge down a vertical staircase filled with boulders and shrouded by vertical rock walls. These horrendous falls start immediately under the Route 123A bridge, which crosses the river at the take-out. Paddlers would do well to familiarize themselves completely with a convenient take-out spot that is sufficiently far upstream from this drop to insure a safe exit.

When the Cold is running high, it is certainly worth one's time to see water blasting through the Drewsville gorge: with large stairstep drops concealed by abrupt rock walls, this gorge is as impressive as any in New England. Even thinking about boating it should be enough to make you wet your pants in fright.

The gage is on the left bank, just upstream from the bridge and the entrance to the Drewsville gorge.

Another nearby river similar in difficulty to the Cold is the Saxtons in Vermont.

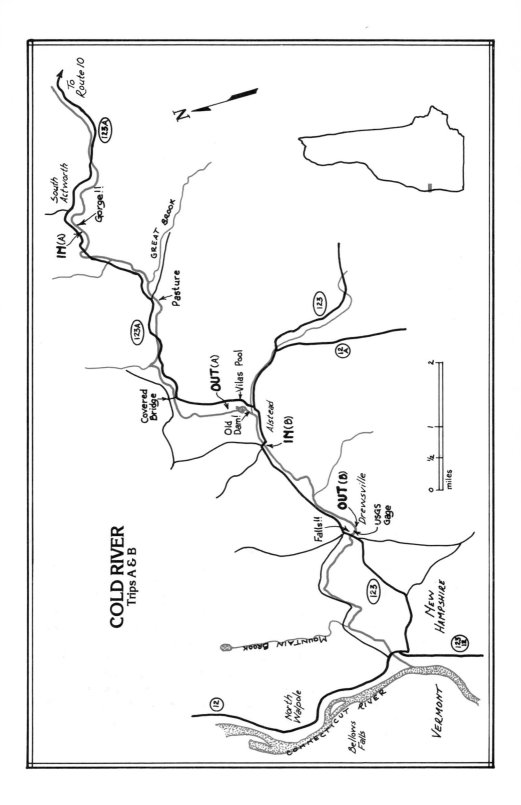

COLD RIVER
Trips A & B

To Route 10

123A

South Acworth

Gorge!!

GREAT BROOK

IN (A)

Pasture

123A

123

12
A

Covered Bridge

OUT (A)

Vilas Pool

Old Dam!

Alstead

IN (b)

OUT (b)

Drewsville

USGS Gage

Falls!!

123

MOUNTAIN BROOK

New Hampshire

123
12

North Walpole

12

CONNECTICUT RIVER

Bellows Falls

VERMONT

N

0   ½   1   2
miles

Contoocook River (NH)

JAFFREY TO PETERBOROUGH
Trip A

Distance (miles)	Average Drop (feet/mile)	Maximum Drop (feet/mile)	Difficulty	Scenery
5	37	100	2	Good

TOO LOW	LOW	MEDIUM	HIGH	TOO HIGH	Gage Location	Shuttle (miles)
		2.8			Peterborough	5

The upper Contoocook is but a fledgling compared to its larger downstream brother. It is a small, easy-flowing stream, which, with one exception, presents a rather mild and consistent personality. This stretch is very narrow at its start, but it does broaden out, and the most distinct rapids occur well after the start of the trip, so one has time to warm up properly. Dropping at a fairly constant rate, this trip presents no special problems other than a sometimes narrow, twisting channel or an occasional tree across the river. The one exception is Gum Drop Rapids, which start innocently enough, but end with a bang that most people would elect to avoid. Ignoring these uncharacteristic rapids for the moment, this is a run for training canoeists and for an easy first-of-the-season trip. The river flows out of a pond which has the effect of regulating extreme variations in level; it should, however, be paddled early in the season to insure enough water. Even then, many parts of the river are less than a foot deep.

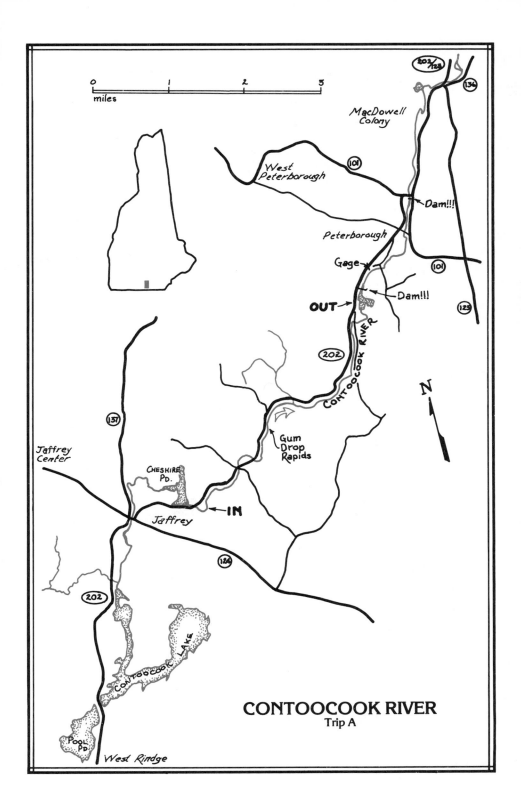

0   1   2   3
miles

MacDowell Colony

West Peterborough

101

Dam!!!

Peterborough

Gage

101

OUT   Dam!!!

125

202

Contoocook River

131

Gum Drop Rapids

N

Jaffrey Center

Cheshire Pd.

Jaffrey   IN

124

202

Contoocook Lake

Pool Pd.

West Rindge

CONTOOCOOK RIVER
Trip A

The upper Contoocook trip starts as water flows over the lip of Cheshire Pond, along Route 202 just north of Jaffrey. Here the river is very narrow and it flows fast through the many trees that populate the stream bed. There are several put-in spots on Route 202 itself as well as one by the Jaffrey sewage pump station plant, located on a side road off Route 202, .5 miles north of Cheshire Pond. By the pump station, the river is 2 to 3 boat lengths wide. Immediately below the station the Contoocook turns right and goes under a railroad bridge. A number of tight turns occur soon after this bridge. Shortly, one sees a factory on the left bank, and the river widens out. Through this whole section and what follows, the paddler should be cautious about trees that may block the river. When the river is away from the road, the scenery is attractive enough, but not spectacular.

The Contoocook continues on as before, passing several more bridges in the process. The paddler should watch for a fairly large island which can be run on either side. There is a house far back on the right, downstream side of the island near a wooden bridge. These are the landmarks for the beginning of Gum Drop Rapids. The island comes after a sweeping left turn; old Route 202 is on the left bank.

As the river approaches the bridge, the rapids increase in difficulty (Class 3 in low to medium water) as sideline rocks narrow the channel. The rapids continue beyond the bridge down to the next right turn (several hundred yards). In this turn, there are two large rocks blocking the center route; there are narrow channels between them, however, and the right turn is sharp and immediate. The outflow is straightforward down to a left turn, where the killer awaits. Here, the Contoocook drops about 6 to 8 feet over a washboard directly into a truck-sized boulder in river center with most of the water then sliding by on the left side. In low water (2.8), the washboard is filled with exposed dragon's teeth, abrupt drops, hydraulics, and it just plain looks nasty. There is another truck-sized boulder directly downstream from the first one. After this boulder, the river regains its senses and levels out in Class 2-3 fashion. Class 2 boaters should avoid Gum Drop by portaging on old Route 202. Others should scout and decide for themselves. Local residents do tell of people who were trapped in the last part of Gum Drop and drowned.

The rest of the river is similar to what precedes Gum Drop except that it may be a little more spirited. When you pass a radio tower on the left bank, the rapids are over, and, in another mile or so, you enter still water behind a mill dam where there is a convenient exit by Route 202. Alternatively, one can portage the dam and run the spicy rapids (Class 3) directly below on into Peterborough. The U.S.G.S. gage is farther down on the left bank, and there is at least one dam in Peterborough, so be wary of still water.

The gage is 100 feet downstream from the mill dam at the end of the trip, 1 mile south of Peterborough.

For a trip of similar difficulty, try the Souhegan.

Contoocook River
(NH)

HILLSBORO TO HENNIKER
Trip B

Distance (miles)	Average Drop (feet/mile)	Maximum Drop (feet/mile)	Difficulty	Scenery
6.2	23	60	3-4	Good

TOO LOW	LOW	MEDIUM	HIGH	TOO HIGH	Gage Location	Shuttle (miles)
5.6	7.4C	9.0C	10.0C	10.0C	Henniker	6

    The Contoocook is one of New Hampshire's largest rivers and offers some of the best heavy water canoeing in all of New England. Because it holds water much better than smaller streams, the Contoocook can be run late in the season and even after heavy rains. In terms of personality, this section is definitely schizophrenic and manic-depressive; it is either disquietingly calm or ravagingly mad, changing almost without warning. Rocks of all shapes and sizes populate this people-eating run, and the level of the water determines how they affect the boating. Low or medium water requires much maneuvering. At higher levels the rocks are responsible for the extreme turbulence. This trip is short: the first part can easily be eliminated and it can be repeated several times in a day. The lower half can be viewed from Route 202 except for two or three places and, as things would have it, these places are Class 4. The scenery is only OK, but you'll never notice.

One starting point is the old Route 202 bridge crossing about 3 miles below Hillsboro. You now have to turn off Route 202 to reach old 202 and this bridge. After a half mile of flat but flowing water below this bridge, the Contoocook turns quietly left, then shortly right. This right turn starts a long section (about 2.8 miles) of heavy turbulent water. This is the more popular of the put-in spots. It is reached from a small dirt road off old Route 202.

In the first rapids, the water funnels to the center and creates haystacks measuring 3 to 4 feet at a gage reading of 9.5. Far from regular, they clutch and slap at the boat, pushing, then pulling, always trying to disorient or overturn it. Looking at this spot in low water, one sees a relatively horizontal riverbed and then, suddenly, just a very slight downward tilt. Put three or four feet of water here and this tilt is amplified like music at a rock concert. A single paddler in a decked boat will be very busy with alignment problems, not to mention the souse holes. At a gage reading of 8.2-8.5, these waves are merely Class 3. After a brief pause of quieter water, there follows another heavy rapids that it is best to run on the left or, if you want the heaviest water, run on the right. The road comes back here, so if the previous rapids have proven too hard, you can use the road as a safe exit. Next up is another goodie, more extended than the previous one, and best run in the center, since rocks complicate the sidelines.

As the Contoocook leaves the road and then prepares to turn right, get ready for the second hardest rapids of the trip—the S-turn (Class 3-4). Composed of irregular waves, souse holes, rocks, and several sudden drops, this rapids should be scouted if you haven't seen it previously. There are several methods of attack; which is the best depends on water level. In high water the left side is the most straightforward, but it is also the most violent and turbulent. There are also several large boulders at the start of the S-turn near the right bank that create some sharp drops. There are some smaller boulders on the left, upstream from the start, and many more everywhere downstream. At lower levels, you must pick and choose among many obstacles. The entire rapids is in the shape of an elongated S, which extends beyond the next right turn and through the next left turn. Overall difficulty decreases near the end. Leave your boat near the top of this rapids and you'll have a long, tough swim. Just ask people who have frequently boated the Contoocook: they have probably seen the S-turn from (in, not on) the water. The start of the S-turn is not easily visible from the road, although the easier lower portions are. After the final left turn of the S, the gaging station appears on the right bank and, following that, there's a schizophrenic turn. The river suddenly becomes calm as Route 202 approaches very close. After several minutes' paddling, the Contoocook lazily turns left, away from the road, then right again.

Just as mythological sirens lured passing sailors with their songs to leave them wrecked on the rocky approaches to their island, so this calm water may tempt the unwary to launch a boat for a pleasant downriver cruise. The river is ever so peaceful-looking here, just like a sleeping vampire. Any who do

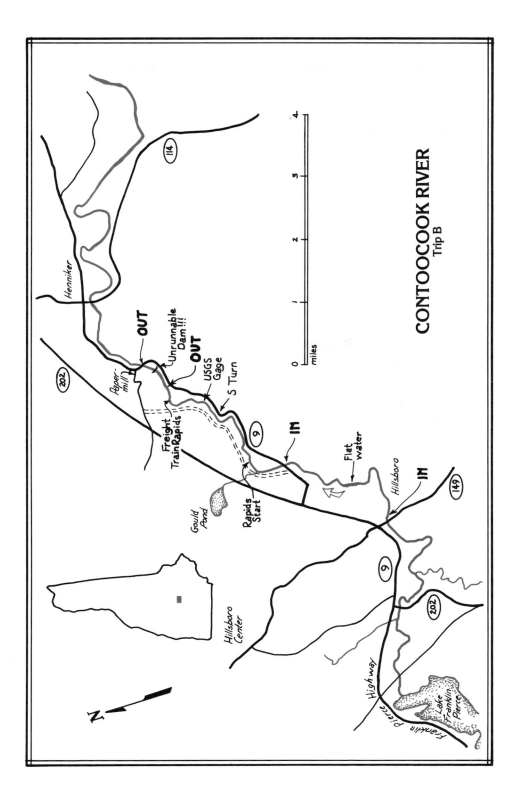

CONTOOCOOK RIVER
Trip B

launch at this point are in for the surprise of the day. Just beyond the right turn is the hardest section on the trip—Freight Train Rapids. In high water this one is a mind-boggler. Large haystacks, souse holes, strong hydraulics, and tricky crosscurrents start quite suddenly and continue for about 300 to 400 yards downstream beyond the next right turn. The difficulty decreases somewhat after this turn. Large boulders line the sidelines all the way down, so don't try to sneak along there or you'll find real trouble with abrupt drops and strong eddies. The best run is in the center; although it has the least security and the most turbulence, it is the safest. This isn't really a contradiction. The run is most difficult at the start and again halfway down. You'll never have a chance to observe this when paddling Freight Train, but it does have a slight S-turn. Freight Train cannot be seen from the road; it is about as far away from Route 202 as the Contoocook gets. On the left bank are a small dirt road and an old railroad bed, but you won't see them from the river unless you are looking. At a gage reading of 9.5 and over, Freight Train is rated hard Class 4 or Class 5 because of its length and turbulence. A swim through it could easily result in serious injury. At a gage reading of 8.5, Freight Train is still Class 4. This trip can end in the calm water just above a dam where Route 202 is once again nearby, or downstream from the dam, at two closely spaced bridges where parking facilities are better. The dam is unrunnable and must be portaged.

The introduction to Freight Train Rapids on the Contoocook — Don Miller

At a gage reading of 8.2 to 8.5, very competent paddlers have taken their own open boats down all these rapids, although it was not easy, and a single mistake could have lost a boat. In general, the Contoocook is not a good open boat river. If a tandem crew must try it, they should at least wait until the gage is below 7.5. At this level there are plenty of rocks to bend your canoe around, but no overpowering waves. At a gage reading of 8.5, the Contoocook is fun for competent closed boaters. At 9.5, it is a bit scary. *At any canoeable level the river's continuous current and width make rescue efforts very difficult for everyone.*

The gage is on the right bank, on old Route 202, about 2.5 miles southwest of Henniker. It is in the Telemark system.

For another river roughly the same size as the Contoocook, try the lower Ashuelot (Ashuelot to Hinsdale). If you want another Class 4 run for the same weekend and you don't want to travel a lot, try the Blackwater, the upper Ashuelot, or the Otter Brook.

Dead River
(ME)

SPENCER STREAM TO WEST FORKS

Distance (miles)	Average Drop (feet/mile)	Maximum Drop (feet/mile)	Difficulty	Scenery
15	28	50	2-3	Excellent

TOO LOW	LOW	MEDIUM	HIGH	TOO HIGH	Gage Location	Shuttle (miles)
	800 CFS	1300 CFS			Flagstaff Dam	20

The Dead River boasts one of the longest wilderness white water trips to be had in New England. By slightly altering the trip described here, a two-day expedition could easily be arranged for those who may prefer that. Despite its reputation, the Dead is not a continuous stretch of difficult rapids, although it does offer some fine canoeing for open boats. Magnificent scenery, typical of the Maine backwoods, is a strong plus for this run. Once you start, however, you have little choice except to complete the trip in one way or another, since the river is isolated from any main roads, although a trail running alongside on the right does appear on topo maps. Because of the isolation and the possibility of quickly changing weather, inexperienced paddlers should carefully evaluate the consequences of an accident before deciding to run.

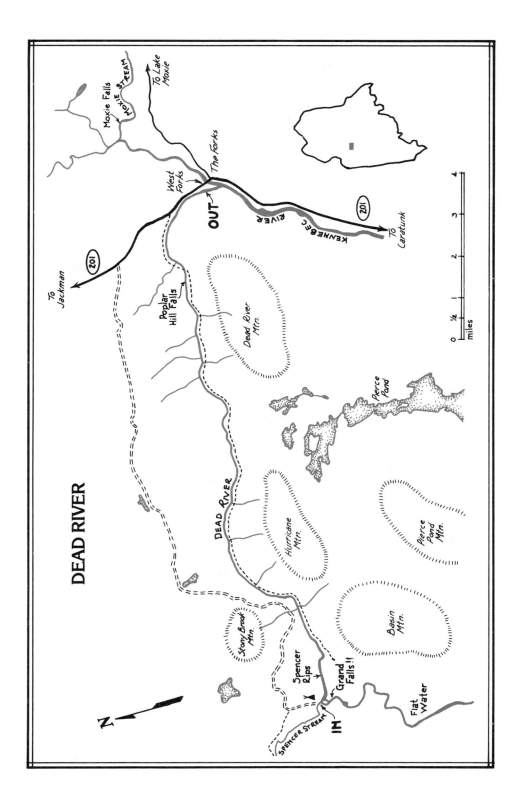

DEAD RIVER

N

To Jackman

201

To Lake Moxie

MOXIE STREAM

Moxie Falls

The Forks

West Forks

OUT

KENNEBEC RIVER

201

To Caratunk

Poplar Hill Falls

Dead River Mtn.

Pierce Pond

DEAD RIVER

Hurricane Mtn.

Pierce Pond Mtn.

Stony Brook Mtn.

Basin Mtn.

Spencer Rips

Grand Falls!!

IN

SPENCER STREAM

Flat Water

0 ½ 1 2 3 4
miles

The Dead has been the scene of the National Open Canoe Championships for a number of years; at race levels the river is primarily Class 2-3. Since the trip is so long, and the rapids are numerous, it is impractical to discuss them in detail, so only a general description of the trip follows. A competent paddler can easily complete the entire course scouting from the boat. Those less sure of themselves may want to eyeball Spencer Rips and Popular Hill Falls.

Arriving at a suitable starting point is as difficult as the run itself. For a put-in at the confluence of Spencer Stream and the Dead, proceed north from West Forks on Route 201 for 3.2 miles and take a dirt road that leads west to the river. This road has been improved in recent years, so it's not the old muffler-scraper it used to be. Directions on which turns to take on this road should be obtained from the locals, since the road changes as the logging route changes. As an alternative to driving yourself, it is possible to contact Ed Webb at Webb's Wilderness Outfitters and General Store in West Forks, Maine (207-663-2214). Other people also do this type of hauling. You can sometimes obtain river release information from these outfitters.

At its confluence with Spencer Stream, the Dead makes a big elbow to the right, and a fair-sized island sits in midstream, blocking the view of Spencer Stream for those who have just come down from Flagstaff Dam. The channel to the island's left is shallow. Just below the island, the Dead turns left and funnels to the right side, cascading over small rocks and standing waves in Spencer Rips. The Rips last for about 300 yards, the last 25 being the heaviest water, with waves reaching two feet at race-level discharge (around 1200 to 1300 CFS). Toward the last of this rapids there is a rock wall extending 20 feet up the right bank, and a big eddy swirls on the left side at the end. At race levels, Spencer Rips is rated a straightforward Class 3. It is a good place to play and is probably the second or third most difficult spot on the trip. It can be run almost anywhere, although most canoeists take it on the right center, since the extreme left is loaded with rocks.

After the Spencer Rips introduction, the next several miles will seem tame, being mostly Class 2-2+. The current moves constantly, and the rocks are small. Notice all along that the larger rocks prefer the sidelines, and it isn't until very high discharges that they really make their presence felt. The Dead then drops into a very large pool-like area which turns left, passing an old wooden logging drop ramp, high on the right bank. Shortly below are some Class 2 rapids, a pool, then some slightly harder rapids with larger rocks and a few 1- to 2-foot drops. A very long stretch of Class 1-2 water follows, where the paddler can relax and enjoy the luscious scenery. For those who like to drift along lazily sponging in nature, this is truly heaven. The entire area is devoid of signs that encroaching bipeds have ever been here, except, of course, for an occasional broken boat or beer can. The conifers point skyward standing at strict attention, contrasting with the gentle swaying of the white birches, while rocky faces protrude from the banks as if trying to break away from their earthly restraints. The end to this section is marked by sharp 100-yard rapids in a fairly straight stretch where, at the bottom right, one large and one small rock point from shore towards a small but powerful hole. An eddy sits slightly downstream near the bank. These rapids are rated Class 2+. They begin a section of drop/pool, drop/pool. After this entering set of rapids there are some good Class 2 and several 2+s, then comes .75 miles of continuous rapids in the Class 3 range. This last tickler has the heaviest water in the middle near the end and an abundance of rocks throughout. Calm water again prevails for a long while, terminating in Class 2 rapids and then Class 2-3 rapids in a right turn.

Upper Popular Hill Falls is located just after a right dogleg. It is a fairly wide rapids, which can be run almost anywhere, although there are usually one or two groups of rocks that necessitate some deviation from a straight course. Upper Popular Hill Falls is rated Class 3 at race levels. Below, the Dead widens only to funnel left into Lower Popular Hill Falls Rapids. The first 100 yards are the heaviest and hardest of the trip (tough Class 3 at 1200 CFS)—at medium levels, it consists of turbulent, fast water rushing among a line of rocks. After this initial stretch, the difficulty gradually tapers off as the river approaches a small bridge in the next turn. From start to finish by the bridge Lower PHF Rapids measures 300 to 400 yards. From this little bridge to the take-out at West Forks behind Webb's Store, the water is Class 1-2. It can be very scratchy if the discharge is less than 1000 CFS. Along the way, you pass a gaging station on the left bank, but there is no outside gage.

The main discharge on the Dead is controlled by the dam on Flagstaff Lake. Because the Dead is dam-controlled, scheduled water releases during the summer make it one of the few rivers that can be paddled then. The dam is also the start of the downriver races. It can be reached via an 80-mile drive from West Forks. The reward for this extra trek is a view of the dam and Long Falls below it, plus the chance to canoe about 6 miles of flat water and portage around Grand Falls. Grand Falls, which drops precipitously about 30 feet, is .25 miles above the confluence with Spencer Stream. It is well worth a visit upstream if you start at Spencer, since the view is spectacular.

Instead of starting at Spencer or Flagstaff, a trip may also be hauled into Fish Pond, which attaches to the northern waters of Spencer Lake. After paddling eastward 6 or 7 miles on these lakes, you reach an outlet dam and, after a rough portage, you can place the boats in Little Spencer Stream. Quite small and serene, this stream drowsily meanders about 4 miles before entering Spencer Stream proper. Two miles beyond, you join the Dead. The road to Fish Pond is Class 3 and turns off Route 201 near Lake Parlin, 15 miles north of The Forks. If the dam at Spencer Lakes is delivering 200 CFS, the trip is fairly enjoyable; much more and the trees would have canoeists for company in the woods.

Fire and camping permits must be obtained for the whole Dead River area through the Forest Ranger in Caratunk, Maine. There are several campsites near the confluence of the Dead and Spencer Stream. Local rangers have been known to get very upset if you go without a permit or 'reinterpret' a permit that you have.

Popular Hill Falls on the Dead River — Bruce Arnold

Deerfield River (MA)

BEAR SWAMP TO ROUTE 2

Distance (miles)	Average Drop (feet/mile)	Maximum Drop (feet/mile)	Difficulty	Scenery
9.5	22	40	2	Good

TOO LOW	LOW	MEDIUM	HIGH	TOO HIGH	Gage Location	Shuttle (miles)
2.6	3.9				Charlemont	9.5

The Deerfield River used to offer a fine Class 2-3 trip through a relatively unencumbered river valley; however, due to the construction of the Bear Swamp Pumped Storage project, this is no longer the case. Built to help meet New England's insatiable demand for power, the project has destroyed some of the best portions of the lower Deerfield, and a good part of the harder, upper section. Although the project does release water, reports are that you should not expect too much cooperation from the personnel at New England Power in so far as advance notice of releases goes, although they will tell you what is being released at the moment. So, you show up and take your chances. There may be a raging torrent, or, on the other hand, there may not be enough water to float a rubber ducky.

To start the trip, put in as close to the dam as you can. The banks are fairly steep, but they make for a good bobsled ride if there is snow. An alternative put-in is near Hoosac Tunnel, about 1.5 miles below the dam. For most of this 1.5 miles, the river is Class 1-2 depending on water level. A large island is just upstream from the railroad bridge at Hoosac Tunnel. Below the island, the river makes a large sweeping left turn. The current can be strong here with occasional standing waves on the outside of the turn. The tunnel is 4.75 miles long, and it is several hundred yards back from the high right bank.

Continuing past a small iron bridge, paddling should be easy unless there is an overabundance of water. You will see several islands and the road intermittently visible on the right bank.

When you notice the riverbed narrowing ahead and a corresponding increase in current, prepare for Zoar Gap Rapids. This is the hardest rapids of the trip—Class 3 in medium water, and several hundred yards long. For most of this distance the riverbed is about half its normal size. Halfway down, a group of rocks blocks most of the central channel, and you need some quick maneuvers to get through. At higher levels, when rocks are covered, the course is straightfoward but turbulent. The outrun from this group of rocks is fairly fast among some large boulders that last beyond a small bridge. The road is close on the right, so you can easily scout on your way to the put-in. This is recommended if you are unfamiliar with the river.

Below the bridge, the river widens again, and there's a picnic area on the left. This is a good starting spot for an easier trip that avoids Zoar Gap. In the next right turn, there are several large rocks which are easy to hit or miss. Farther along is a hundred-yard stretch where the rocks are denser still, technical Class 2 and as hard as you find in this section. Rocks are as big as large bread baskets, or perhaps even oven-sized. From here to the take-out near Route 2, the Deerfield is uncomplicated, with just an occasional ripple or rock in a large riverbed.

Immediately past a fair-sized island, the Cold River enters from the right with accompanying rapids. A railroad bridge follows, then Route 2 shows up overhead. Take out below the bridge by a roadside park on the left. Approaching Route 2 can be very scratchy or annoying in low water—colorful seaman's language has been overheard there. Cars may be left at the roadside park, but alas, there are no water closets, which are so important after a long trip in a tight wetsuit.

To reach Bear Swamp, go west out of Charlemont on Route 2 until you spot a bridge across the Deerfield. Turn right just before this bridge. Follow the main road until you cross the river just below Zoar Gap Rapids. Then continue along the right side of the river until you see the dam. A bit before crossing the river, you will pass through a narrow underpass where there are two sharp, blind turns.

The gage on the Deerfield is in Franklin County, on the left bank, one mile downstream from Charlemont and 2.5 miles downstream from the

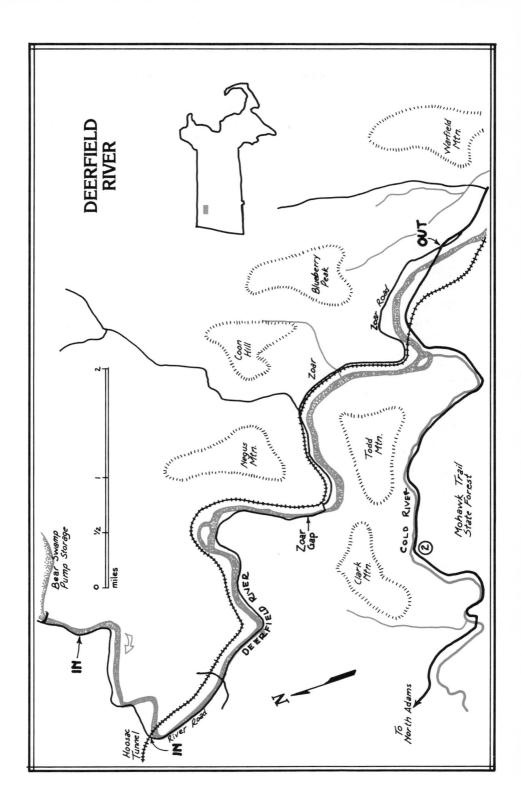

DEERFIELD
RIVER

mouth of the Chickley. You can easily see the concrete gage house from Route 2. Do not rely too heavily on this gage: the important water is controlled from Bear Swamp. You should also be aware that the gage is located in a very wide section of the river, so a one-foot rise here will usually mean a much larger increase in the level at Zoar Gap.

Canoeability ratings given refer principally to Zoar Gap. The HIGH reading doesn't mean that there's real big water, only that Zoar Gap is mostly washed out by a strong current. If you would like another run that is nearby, although harder, try the Chickley. For another run of Class 1-2 difficulty, try the North River.

Approaching Zoar Gap Rapids on the Deerfield at relatively high water — Debbie Arnold

Ellis River
(NH)

ROUTE 16 TO HARVARD CABIN

Distance (miles)	Average Drop (feet/mile)	Maximum Drop (feet/mile)	Difficulty	Scenery
3.0	87	100+	4	Good

TOO LOW	LOW	MEDIUM	HIGH	TOO HIGH	Gage Location	Shuttle (miles)
1.3C	2.1C	2.9C			Route 16	3.0

The Ellis River typifies the White Mountains. Starting near the slopes of Mount Washington, it cascades down the very heart of Pinkham Notch, contorting among rock formations and rapids to drop finally into the Saco River near the junction of Routes 16 and 302. In its upper part the Ellis displays some of the most tempting but impassable rapids anywhere in New England. It's intriguing to scout these sections, wondering how you would run the blind 90 degree turns, or squeeze through the narrow slots when the water itself has trouble doing so, but these rapids are for looking only. Lower down rapids become more manageable, with stretches of calm water and some variation in sophistication and difficulty. This is the section for sport, not stunt.

The Ellis is relatively small with a watershed to match. It seldom has enough water for running, but, when the snow melts and the white water eggs hatch, it does offer a fine adventure. It is interesting to note that the Ellis can be run with as little flow as 129 CFS (gage of 2.1). A modest flow of 350 CFS means water a foot deeper (gage of 3.0).

Since Route 16 parallels the river, you can start a trip from any number of places. One spot that kicks off with a Class 4 bang (Commencement Rapids) is about 0.4 miles above a small bridge on Route 16. The bridge is 0.3 miles north of Dana Place. If you are heading north, the river crosses from the right to the left side of the road at this bridge.

At the put-in there is a small roadside turnoff on the east side which is handy for changing into wetsuits. The trek to the river is short and steep. The river here looks as if it has more rocks than water. Finding a calm spot to launch a boat and fasten your spray skirt is not trivial. Above, the channel bounces down a relatively narrow slot that seems nearly impassable (it isn't at low water). Below, the rapids are uninterrupted and technical and usually there is only one practical route through the labyrinth. Water level is crucial for this stretch. At a gage reading of 2.1, the going is a tight Class 3. Higher levels will most certainly be Class 4 or more. At any level the boater must make one quick decision after another to negotiate the maze successfully: the correct path is not obvious; the current is pushy, and broaching opportunities abound. If this section appears too hard, skip it. Do it at the beginning of a second run. Also, open boats had best forego Commencement Rapids. At low levels good open boaters could handle most of the other rapids.

At the Route 16 bridge, 0.4 miles below the put-in, the furious pace slackens, and a section of more moderate rapids begins. First comes a relatively straightfoward chute, near a contemporary house on the left bank. Rapids then alternate with calmer water for the remainder of the trip. The unbroken difficulty of Commencement Rapids is behind you. From now on the Ellis's course is generally away from Route 16.

There is one drop through a narrow rock gorge that should be portaged. This cataract, about 25 yards in length, requires that you negotiate several stairstep drops and hydraulics while placing your boat in the middle of the only runnable channel. Each boater should judge for himself whether to attempt a run. The walk around is easiest to manage on the left. Preceding this drop is a left turn, a short stretch of calm water, then a right turn into the falls. This one is Class 5, even in low water.

Another noteworthy rapids occurs where a rock ledge blocks the river. In low water, the channel funnels to the left into the only runnable path. At higher levels, the water will pass more directly over the ledge, opening up new routes and new problems, but that's the fun of it all.

One convenient take-out is near the Appalachian Mountain Club's Harvard Cabin. The cabin cannot be seen from the river; a large summer home near the river on the left bank is a landmark. The house is close to a 1- to 2-foot stairstep drop into a pool. Ask permission if you plan to cross the land. Harvard

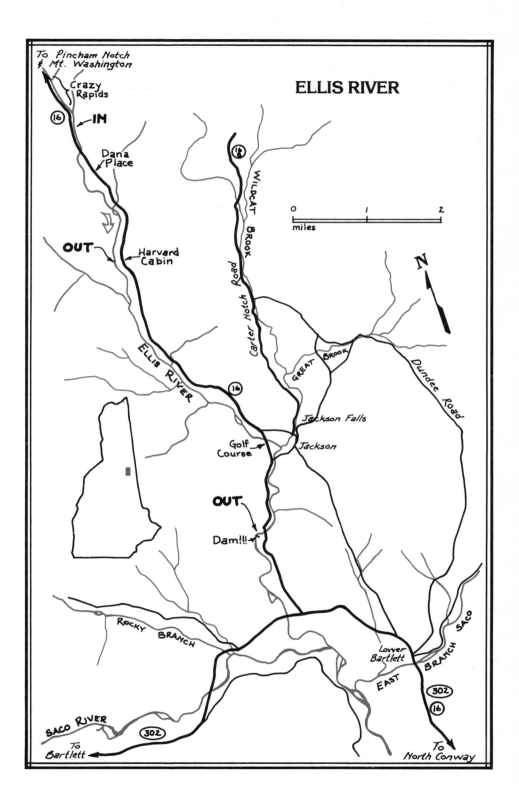

ELLIS RIVER

To Pincham Notch & Mt. Washington

Crazy Rapids

16 IN

Dana Place

18

Wildcat Brook

Carter Notch Road

OUT

Harvard Cabin

Ellis River

16

0   1   2
miles

N

Great Brook

Dundee Road

Jackson Falls

Golf Course

Jackson

OUT

Dam!!!

Rocky Branch

Lower Bartlett

Saco

East Branch

302

16

Saco River

302

To Bartlett

To North Conway

Cabin is 6.1 miles up Route 16 from the junction of Routes 16 and 302. A small dirt road off Route 16 leads to the cabin (about 100 yards).

From Harvard Cabin the Ellis continues for 5 more miles to the dam at Goodrich Falls. Parts of this section are flat and uninteresting, but there are rapids, some of which pack a wallop in high water. You'll find nothing on a par with Commencement Rapids, but there are some which you should scout, particularly one set where the river seems to disappear over some rocks. Another obstacle to note in this lower section is a golf course just north of Jackson. Flying balls and golf carts are definitely extraordinary white water hazards, but you might have to deal with them here. Consider yourself "Fore" warned. The dam at Goodrich Falls is just after a Route 16 highway bridge; the actual drop looks as if it plunges straight into hell.

The gage is in Carroll County, on the right bank, 1.3 miles upstream from the bridge near the put-in.

Farmington River (MA)

OTIS BRIDGE TO ROUTE 8
Trip A

Distance (miles)	Average Drop (feet/mile)	Maximum Drop (feet/mile)	Difficulty	Scenery
2.4	27	50	2	Good

TOO LOW	LOW	MEDIUM	HIGH	TOO HIGH	Gage Location	Shuttle (miles)
	4.0	5.0			New Boston	2.4

Compared to portions farther downstream, this section of the Farmington is small and fairly slow-moving at low levels, with few rocks. Run either as a separate trip or as a warmup for the more exciting section below, it offers only a minimal challenge to the competent, but it is good water for the novice. Route 8 passes alongside the entire length, so an exit anywhere is relatively easy.

Put in at a bridge that leads to Otis Reservoir from Route 8, or farther up river toward Otis for a longer trip. At the bridge, the Farmington is 50 to 75 feet wide and smooth-flowing. After a couple of left turns, a few rocks appear with some good-sized ones in the next turn. A nice pool follows. Then come some rocky Class 2 rapids, a brief pause, then another rapids, then calmer water. The river next loops to the right, around a picnic area and a camp on the right bank. Water here is Class 1-2. Downstream, after the river approaches Route 8 and

143

turns left, lie more rapids. The rocks start sparsely but increase in frequency; several large ones are located at the bottom on either side. There's a small drop and a hydraulic in the center. The gradient is steep but the run is not too difficult. After a short breather, a similar though somewhat more strung out rapids appears. At the end some large rocks extend from the right bank, so stay left where there is an avenue through several more rocks. In medium water all of these will be mostly covered, allowing a straighter course. Along the way, the left bank is heavily wooded and the right is fairly steep, leading to Route 8.

For the gage location, see the next trip.

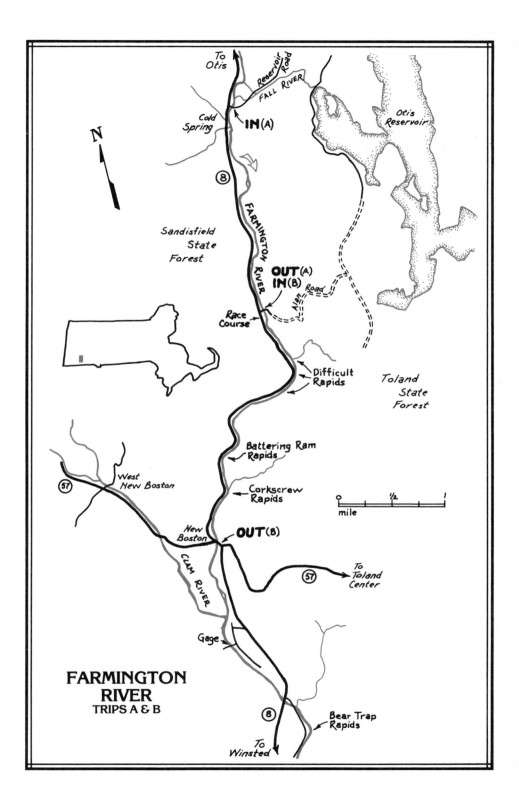

N

To
Otis

Reservoir Road

FALL RIVER

Cold
Spring

IN (A)

Otis
Reservoir

⑧

FARMINGTON RIVER

Sandisfield
State
Forest

OUT (A)
IN (B)

Alan Road

Race
Course

Difficult
Rapids

Toland
State
Forest

Battering Ram
Rapids

West
New Boston

⑤⑦

Corkscrew
Rapids

0   ½   1
mile

New
Boston

OUT (B)

To
Toland
Center

⑤⑦

CLAM RIVER

Gage

FARMINGTON
RIVER
TRIPS A & B

⑧

Bear Trap
Rapids

To
Winsted

Farmington River (MA)

ROUTE 8 TO NEW BOSTON
Trip B

Distance (miles)	Average Drop (feet/mile)	Maximum Drop (feet/mile)	Difficulty	Scenery
3.0	75	100	3-4	Fair

TOO LOW	LOW	MEDIUM	HIGH	TOO HIGH	Gage Location	Shuttle (miles)
3.6	4.0C	4.5C			New Boston	3.0

The Farmington River above New Boston is a fine Class 3-4 run provided that there is enough water. The minimum level is critical. Three hundred CFS is adequate for a relatively fluid run, although the river is still passable at 250 CFS (with much bottom-scraping and cursing). In general, lower levels require more maneuvering, and certain passages even become dry or unrunnable. Except for the spring runoff, the river is usually too low for canoeing, although the fall (October, most of the time) water releases from Otis Reservoir via Fall River raise the level sufficiently for a trip. The river is very fast-flowing, with essentially no flat water and many rocks of all shapes and sizes. In several locations there is only one canoeable channel, so the party should stay well spaced to avoid a chain-reaction pileup in case of trouble. Route 8 follows the river, and most of the rapids can be scouted from the road.

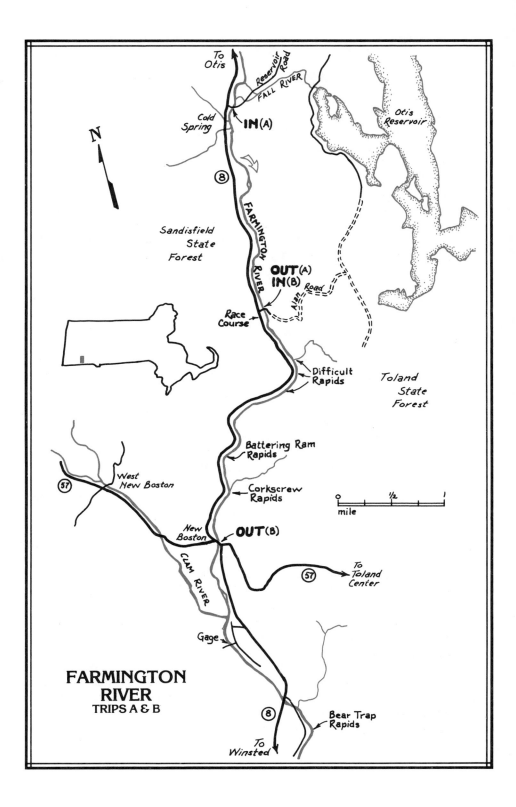

N

To
Otis

Reservoir Road

FALL RIVER

Cold
Spring

IN(A)

Otis
Reservoir

8

FARMINGTON RIVER

Sandisfield
State
Forest

OUT(A)
IN(B)

Race
Course

Road

Difficult
Rapids

Toland
State
Forest

Battering Ram
Rapids

West
New Boston

57

Corkscrew
Rapids

½   1

mile

New
Boston

OUT(B)

CLAM RIVER

57

To
Toland
Center

Gage

8

Bear Trap
Rapids

FARMINGTON
RIVER
TRIPS A & B

To
Winsted

A good meeting spot for this trip is near an iron bridge, just off Route 8, 5.2 miles south of Otis. After putting in on the upstream side of this bridge the paddler passes immediately into the area where an annual slalom race takes place. The river then bears right, passes a picnic area on the right shore, and then turns left again. A hundred yards or so downstream is an abrupt drop of two feet, best run on the right or left, with the center reserved for high water only.

The outflow is straight, and the eddies are plentiful. This is the first in a series of closely spaced rapids in the upper section of this run. The Farmington continues to flow swiftly among rocks that clutch and grab at the canoe. Just before and on into the next right turn are some rapids best run on the extreme right, especially in low water. More adventurous boaters could start out on the extreme left and pass diagonally to the right halfway down, in front of several large rocks.

The next rapids appears after a downstream passage of several hundred yards amid partly covered rocks and backcurlers. A center cluster of large rocks guards a drop over a ledge which it is best to run on the right. There are eddies below this rapids, then the river races toward the next drop — Decoration Rock Rapids — 100 yards downstream.

Decoration Rock Rapids is very intricate in low water, requiring you to execute several sharp turns in quick succession. Look this one over briefly beforehand if it isn't familiar. Approach in the left center and stay on the lookout for small rocks. Snake your way down and pass a ten-foot-high rock on the left bank. Keep it to your left, and then move to a central channel to avoid another rock below. Mess up one of these short turns and you will find yourself an ornament on Decoration Rock itself. It is possible to pass to the extreme right, but careful rock-picking is necessary, i.e., picking the rock on which you want to broach. A boulder patch upstream divides the two courses. At 300 CFS the left center route is a straighter shot since the small, menacing obstacles are almost covered. At a gage reading of 4.5, both routes are fairly straightforward and Decoration Rock supports a large upstream pillow.

The next right turn holds another drop over a ledge, best run to the extreme left of center. Several hundred yards below, where a group of rocks stands on the left side, is one of the best playing holes on the trip. A slanting 2- to 3-foot drop into a hydraulic is easy to punch through, but try coming back and sitting in it sideways. From this spot onward, the river's rapid flow over and around rocks creates standing waves and hydraulics. The general pace will be Class 3 or a little harder. The gradient is not so steep as that of the first half mile, but an afternoon sun will shine directly into your eyes.

Downstream, the river turns right, runs past a road turnoff on the right shore, approaches Route 8, then turns left again. Shortly after this left turn, the channel narrows and there is an abrupt drop over a 2- to 3-foot ledge. The current runs diagonally, right to left. This is Battering Ram Rapids; your best approach is right center. This rapids got its name from an old tree trunk that formerly protruded into the canoeist's path. Several people tried to move the tree while running the drop. One might say that they got out on a limb.

After Battering Ram Rapids, the river angles away from the road, turns right again, sweeps through another rock garden, then divides around a small island. Past the island on the left are the remains of an old concrete bridge and a stone wall. Below these landmarks of times past, the river turns right and drops over two sharp ledges. After another left turn, the approach to the last big rapids — Corkscrew — begins.

If the water conditions allow it, a center approach to Corkscrew is best because it provides options of moving either left or right at the end. The serpentine route ends in a 2- to 3-foot drop, angling right to left, with an outrun that is a bit more complicated than it first appears: several angled hydraulics can catch an unwary paddler who thinks the worst is over. At a gage of 4.5, Corkscrew becomes harder than at lower levels. The left side opens up as a viable route, but there are some gasping souse holes just waiting to swallow nice, shiny new boats. Look it over first to decide if you want to run. The center route at 4.5 is considered the toughest path. If you botch the job and dump, there is usually a full complement of people sitting on the bank to applaud your lack of finesse. At a gage reading of 4.5, Corkscrew is a short but intense Class 4. You can scout both this drop and Battering Ram Rapids from the road. For paddlers who have survived this far, the last half mile to New Boston should cause no great concern, although there are still plenty of rocks.

The take-out at the bridge in New Boston has become something of a problem due to local residents and traffic congestion, especially during October when annual water releases bring hordes of boaters out for a good time. In the past, these conflicts have been resolved, but boaters should remember that local property owners have a strong influence on our sport since they can restrict river access. One solution to this problem is to show consideration for the locals, to respect their wishes and their property. Arguments ultimately end in a "no win" situation.

Even though this trip supposedly finishes in the town called New Boston, a post office at the take-out displays a sign that reads Sandisfield, Massachusetts. The local hostelry, on the other hand, is called the New Boston Inn. Send cards or letters if you can figure this one out.

At 300 CFS, this trip is somewhat scratchy and annoying. Channels are narrow and many rocks acquire new color as boats bounce over them. At 600 CFS (gage of 4.5), the whole run is considerably more fluid, though many rocks are still showing and are therefore the target of abuse. The trip is best for covered boats, although the skilled could manage open boats, either singly or in tandem, even at 600 CFS. But should you make a mistake, remember that the Farmington is somewhat unforgiving: a paddler-less boat, will probably broach on one of the many rocks. Consider yourself warned.

The gage is in Berkshire County, on the left bank, five feet downstream from a bridge, 0.3 miles downstream from Clam River and one mile south of

New Boston. To reach this spot, go south on Route 8 and turn right onto an old paved road. A half-mile ride takes you to the bridge. This gage is part of the Telemark system, so the Corps of Engineers can phone to determine the water level.

For another river of similar difficulty in the same general vicinity, try the Sandy in Connecticut.

The Farmington River with a level characteristic of the fall release — L. Papp

Farmington River (CT)

RIVERTON TO ROUTE 44
Trip C

Distance (miles)	Average Drop (feet/mile)	Maximum Drop (feet/mile)	Difficulty	Scenery
12.4	14		1-2	Fair

The most noteworthy portions of this trip are at its extremes. At the beginning is the famous Hitchcock Furniture Company, and at the finish is an area known as Satan's Kingdom. The rapids in Satan's Kingdom, undoubtedly the hardest on the trip, are not characteristic of it. Typical rapids on this trip are Class 2, with occasional channels branching off the main course that offer a chance to meander quietly down small waterways in peaceful serenity.

At the put-in, the Farmington is very wide and slow-flowing with a smooth current. A roadside turnoff by the Route 20 bridge crossing is handy for unloading boats. A mile or so downstream, a boulder patch necessitates some easy rock-picking. From this point to the Route 44 bridge, canoeing is placid. Along the way, small channels leave the main river and the canoeist has a choice of avenues.

Under the Route 44 bridge, small rapids challenge the novice. A down-the-middle run is fine. About 500 yards below is the drop into Satan's Kingdom (Class 2 to 3 at summer levels). Before the Route 44 bridge, on the right, is a roadside turnoff for those who don't want to run the rapids in Satan's

Kingdom. At the drop itself, with the banks towering above the paddler, the illusion is of a short gorge. The river is rock-lined on either side, and the riverbed narrows to tumble over some large midstream boulders. At low or medium levels, the easiest approach is to the right of center. A somewhat trickier approach is on the extreme left. Large rocks at the bottom divide the two routes, and the final leg of the left channel is just large enough to fit a canoe. Another smaller rock, slightly upstream from these dividing boulders, may force another maneuver depending on the water level. The inexperienced should scout this rapids before running. It is no more than 10 to 20 yards long, and a pool below is good for recovery. In high water it can become rather fierce. This is the only rapids in Satan's Kingdom. Afterwards, the Farmington broadens out again and continues flowing quietly with easy rapids interspaced. Take out anywhere, since Route 44 follows the left bank.

There is a U.S.G.S. gage near the put-in at Riverton but it has no external staff, so it is useless to canoeists. It is worth mentioning that your can sometimes run this trip during the summer.

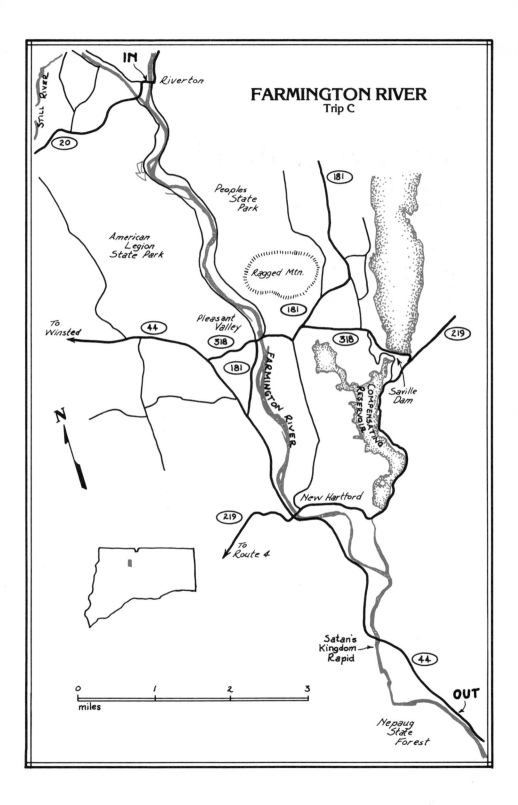

FARMINGTON RIVER
Trip C

Farmington River (CT)

ROUTE 189 TO ROUTE 187
Trip D (Tariffville Section)

Distance (miles)	Average Drop (feet/mile)	Maximum Drop (feet/mile)	Difficulty	Scenery
1.5	15	40	2-3	Fair

TOO LOW	LOW	MEDIUM	HIGH	TOO HIGH	Gage Location	Shuttle (miles)
	1.9	3.2C			Tariffville	2

Every club runs a section of river ominously known as "The Gorge", and the Tariffville Gorge is probably the best known in the western Massachusetts-Connecticut area. This section of 1½ miles is easy to do several times in one day, and it presents several opportunities for playing in the rapids. In low water it is a good place to train beginners. A slalom has been held here in the spring. Low water levels show the most rocks, whereas medium levels generally have haystacks and a fast current.

A put-in is possible near the Route 189 bridge or, instead, follow the main road through Tariffville to a "Dead End" sign, bear left and continue onto a dirt road on the right that eventually leads to a sewage treatment plant, a field used for camping, and then to the river itself. Local construction, however, may eliminate this as a camping site and put-in spot. At either spot, the river is very

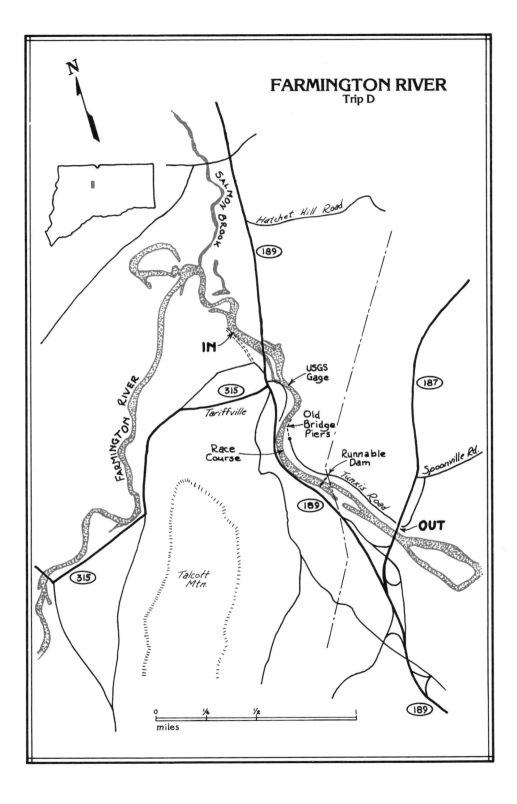

quiet and slow-flowing. Slightly downstream from the upper starting point are the remaining pillars of an old railroad bridge, and beyond, as the river turns left, there is a 6-inch drop that extends across the river at low water. The Route 189 bridge then crosses the river. About .25 miles below, after a right turn, the boater can see a concrete wall on the right and the nearly vertical walls of the so-called gorge forming in the distance. Just inside the entrance to this canyon section, you can read a government gage on the right bank. Slightly below the gage is a right turn with a small ledge. Shortly after this right turn, the first extended rapids appear in which there are two sets of four concrete blocks each, in the right center and left. The blocks are the remains of a bridge that connected Tunxis Road on both sides of the river. Both sets line up parallel to the current. The most exciting ride here is on the extreme right side, although other passages are certainly possible. There is a small pool below on the right.

The Farmington next makes a left turn as the current quickens. The area surrounding this turn is the scene of the slalom. Immediately after the river straightens out, there is a rapids consisting of two sets of two ledges each. The first ledges, located mainly on the right, are about 15 feet apart and drop about 1 foot each. Several rocks populate the middle. When they are covered, in medium water, they can create wicked holes. Ledges in the next series are about 15 to 20 feet apart. They begin about 20 yards below the last ledge in the first set. Each drop is 1.5 to 2 feet. Standing waves and hydraulics greet the paddler before and after each ledge. The worst can be avoided by pursuing a path on the extreme left—this is the usual route for open boats. In medium water, the drops are not so pronounced, but the water is more violent and the current is very fast. As the water level increases, the hydraulics at the end of the last ledge become *extremely mean.*

A few seconds' paddle reaches the next rapids. The entrance is guarded right and left by a line of rocks that extend centerward, but they show up only in low water, so plot an initial central route. You then have the option of bearing diagonally right or left. Rocks in the middle force a decision, and he who hesitates is wet. The left course is probably a little easier but has an S-turn outrun that requires several sharp turns. The right-side path ends in an abrupt drop over a ledge near the right bank. In medium water, this rapids is replaced by a series of small ledges and standing waves with the main current angling to the right. The only exposed rocks are a group in the left center at the end. These rapids are worth scouting.

Below looms a broken dam and the decison of whether to run or not to run. The dam is concrete and it extends from the right bank. There is a 25-foot gap on the left side through which the entire river pours. It is easy for closed boats to run in low water, if you don't mind banging your stern. For open boats, the 3- to 4-foot drop into a cauldron of boiling water is somewhat trickier, but still possible. Solo paddlers have a better chance. Submerged obstacles of an unknown variety have been found at the bottom of the drop. Below the dam, the outflow pours into a car-sized rock that supports a large pillow when the water is high.

At medium levels, the turbulence below the dam flows up and over the car-rock, and an ugly hydraulic extends diagonally downstream from the right side of the break in the dam. Also, the drop itself is less abrupt at higher water levels. A large pool sits below the dam, and there is an outlet on the left side where there is a fast current and a sharp right into a narrow channel. This channel is a way around two islands, and the current can be strong, as can the eddies and crosscurrents. Once past this secton, you will see the Route 187 bridge.

You can reach Tunxis Road by turning off Route 187 onto Spoonville Road, which is just north of the Route 187 bridge crossing. Tunxis Road runs alongside the river on its left side, but only from the race course down. It is not visible from the river.

If you want the gage reading before put-in, the gage is on the right bank, 0.3 miles downstream from Route 189, behind a house at Tunxis Road (west side of the river). It is at the bottom of a steep hill where trash is sometimes dropped.

Looking downstream at the broken dam. The drop is some three to four feet and Car Rock can be seen directly below. A large piece of the dam is immediately downstream of Car Rock — Ray Gabler

Gale River
(NH)

FRANCONIA TO AMMONOOSUC RIVER

Distance (miles)	Average Drop (feet/mile)	Maximum Drop (feet/mile)	Difficulty	Scenery
7.5	25	80	1-4	Excellent

TOO LOW	LOW	MEDIUM	HIGH	TOO HIGH	Gage Location	Shuttle (miles)
0.5		1.0C 1.0			Franconia	5.5

The Gale River is one of New Hampshire's finest. It has everything that counts on a white water trip — in abundance. Isolation, natural beauty, distinctive rapids, and even a gorge all combine to make this an unforgettable voyage. The Gale even has a most picturesque spot for eating lunch or watching skinny dippers. But the most remarkable thing about the Gale is that it is seldom boated. The parts that can be seen from the road are rather dull and uninteresting and would escape the attention of a casual observer. Like an iceberg, the important parts of the Gale are hidden: varied and nearly uninterrupted rapids. If that weren't enough, the Gale's rapids progress in sequence from Class 1 to Class 2 to Class 3 and end with a very definite set of Class 4. Except for the Gorge section, which can be portaged, although with difficulty, open boaters could handle all rapids — and have a good time at it — provided they are adept at maneuvering.

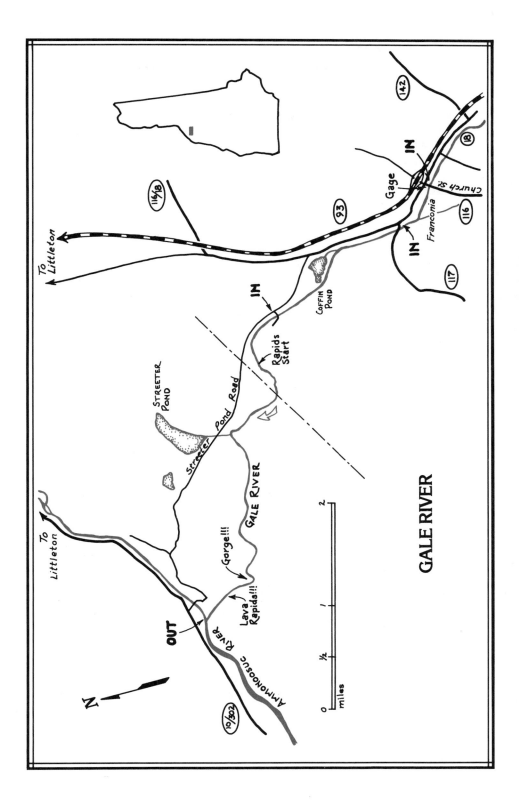

GALE RIVER

Draining parts of Mount Lafayette and Mount Garfield, the Gale flows quietly through Franconia, New Hampshire, displaying only Class 1-2 difficulty. You can start from several different bridges depending on the amount of warmup you desire. At a gage reading of 1.0, the depth here is 3 to 12 inches, depending on where you stick your paddle. If you put in by Church Street, you get to pass an old iron kiln on the left bank that has been standing there since the nineteenth century. A road marker on Route 18 opposite discusses its history. After the kiln, Route 116 crosses the river. From Church Street to Streeter Pond Road Bridge (2.5 miles), the river is mostly Class 1 and shallow, with fairly unattractive scenery, but this is the admission price you pay to the river gods for the good times ahead.

Within a half mile of the Streeter Pond Road Bridge, the paddler turns left and enters a rather isolated valley surrounded by hills which come into view before the turn. After the next right turn, an overhead power line is the last sign of civilization you'll see until you approach the bridge at the Ammonoosuc at the trip's end. From here on the river runs through woods; the current is uninterrupted, and the rapids nearly so. At the beginning rapids are Class 2 with distinct stretches of calmer water in between; then distinctions begin to vanish as the rapids pack closer and closer together. At a gage of 1.0, most rapids toward the beginning of the trip are Class 2, but they are technical and require some maneuvering.

As the rapids increase in intensity and frequency, be especially careful of a particularly deceitful one. This little devil appears to be a straight chute in river center between two boulders in left and right center. It has an abrupt drop of about 2 feet and falls onto a flat rock directly downstream from the drop. The flat rock is difficult to see from upstream and unwary paddlers will blast right into it. An angled run to either side avoids the problem. Stop and scout if you recognize this spot beforehand; otherwise, say hello to the rock.

The Gale continues its descent, and by now it should be obvious that the gradient is picking up, as are the waves. Continuous Class 2 and 3 rapids (gage of 1.0) are common. Rocks everywhere make maneuvering a must. For competent boaters, this section is a playground. You can dart in and out almost at will, catching eddies and hydraulics, and skirting rocks in various ways. In this section the unskilled will find several opportunities to redesign the lines of their boats. Open boaters will be busy but not overwhelmed in this section, although the rapids are continuous. The scenery, if nothing else, is worth the trip. The next thing to look out for is a huge rock outcropping on the left bank, followed by an S-turn (right turn, then left). This is the first warning of the upcoming gorge section, where the difficulty increases instantaneously.

As you complete the left turn of the S, look downstream for a large rock outcropping jutting out from the right bank at a point where the river continues to turn left underneath and then appears to double back on itself. The Gorge (Class 4) is imminent, so pull over immediately to scout. The Gorge is narrow and rock-lined, with nearly vertical sides. There are three major drops.

Fortunately, there are well-located eddies where you can rest and reconsider your previous decisions to run.

The first drop is directly under the signal rock. It is a river-wide ledge, steepest on the right (3-foot drop at a gage of 1.0). It becomes less steep as you go left, but the danger there is that the fast current could mash you against the solid rock wall of the left bank as the river bends sharply right. At lower levels, this difficulty may not be quite so tough. Several smaller drops upstream determine your approach. The hydraulic at the bottom of the ledge can be strong.

After the right turn, a series of rocks across the river creates the second drop. Water level will determine your route. At medium levels, a course on the extreme left opens up, and a trickier path in the middle still remains.

The third drop follows the second drop closely. It is another river-wide ledge, most abrupt on the left, but with a relatively smooth tongue in the right center. At medium levels, watch out for the vicious-looking hydraulic on the left side. Standing waves form the outrun. Several smaller drops downstream should cause no concern. The Gorge veers right into more open surroundings — the steep walls fall back into the woods. There is no pool at the end of the Gorge as you might expect or hope; there is a short stretch of Class 1 to 2 water. At a gage reading of 1.0, the Gorge is for closed boats only, and then only after careful scouting.

The rapids that follow the Gorge are Class 2 to 3 depending on the level. A small, wooden aqueduct on the right bank signals the approach of Lava Dam Rapids (Class 4). The last major rapids of the trip, Lava drops between 5 and 6 feet through a break in an old lava flow. Rocks and hydraulics clutter the approach, but the drop itself is rather straightforward, or so it seems until you try to run it. An angled wave running diagonally down the drop masks some rocks below it; this wave will also threaten to flip any boat not set up properly. The outflow from the drop is fast; the swim is longer than you want to take; and, there are several important rocks to avoid. After Lava Rapids, it's a short distance to the confluence with the Ammonoosuc River where, as at the start of the trip, the Gale wears a Class 1-2 facade.

There is a hand-painted gage on the Gale, under the Church Street Bridge in Franconia on the left side of the river. The canoeability rating of 1.0 is MEDIUM for an open boat on all sections but the Gorge and Lava Rapids. This level is fun for closed boats also, and the Gorge is rated MEDIUM for them at 1.0. Any prospective boater should also realize that even though the Gale River valley is excellent scenery, it is isolated. Help in an emergency will be difficult to find.

Green River
(VT-MA)

GREEN RIVER TO WEST LEYDEN
Trip A

Distance (miles)	Average Drop (feet/mile)	Maximum Drop (feet/mile)	Difficulty	Scenery
6.8	30	50	2-3	Excellent

TOO LOW	LOW	MEDIUM	HIGH	TOO HIGH	Gage Location	Shuttle (miles)
2.8	3.9				East Colrain	6.5

The Green River is known more for its rustic scenery than for its challenging rapids. Snaking its way through the Vermont woodlands, the Green River valley typifies the New England outback. It's a wintry fairyland after a freshly fallen snow, with the white-covered limbs bending down trying to touch the water as it dances between snow-capped rocks. Springtime radiates all the wonder of rebirth in this narrow valley. If the weather is clear, an unchallenged serenity prevails and you see, smell, and hear the woods reviving after a long sleep. The rapids themselves are a mixture of standing waves with an occasional rock garden. They are pleasant but not difficult.

At Green River, Vermont, the river is barely 25 feet wide; a trip may be launched just below the dam and covered bridge there. A trip may also be started several miles upstream where the river is still narrower, with a great deal

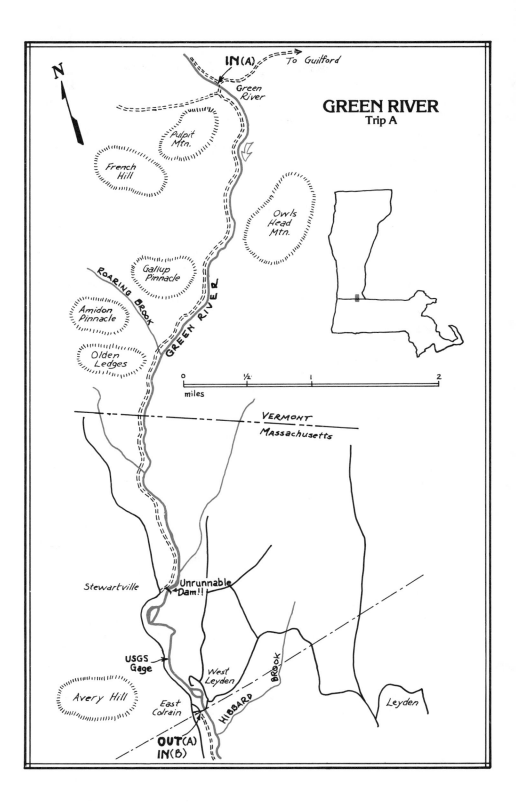

N

IN(A) To Guilford

Green
River

GREEN RIVER
Trip A

Pulpit
Mtn.

French
Hill

Owls
Head
Mtn.

ROARING BROOK

Gallup
Pinnacle

GREEN RIVER

Amidon
Pinnacle

Olden
Ledges

0   ½   1   2
miles

VERMONT
Massachusetts

Stewartville   Unrunnable
Dam!!

USGS
Gage

Avery Hill

West
Leyden

HIBBARD BROOK

Leyden

East
Colrain

OUT(A)
IN(B)

of fallen trees and brush. Although it is broken in at least one spot, a 4-foot dam about .5 miles north of the covered bridge must be carried. A dirt road (Class 2+ if it's wet) runs alongside the river for its entire length, crossing at Green River, Vermont via a covered bridge. (There is a fine of $2 if you drive across the bridge at more than a hiker's pace.) The scattering of rocks and steady current at the put-in are typical of this small stream. The rapids should not challenge a competent canoeist. Water level is critical, and it is usually too low for an enjoyable trip. Only the spring runoff and very heavy rains raise the level sufficiently to offer good sport. The Green is best for open boats.

A third of a mile below the covered bridge, just prior to a right turn, is a small island dividing the river; then the river bends left. One hundred yards below, some snappy rapids appear; they should be easy to run. The river then loops away from the road. When it comes back, the Green narrows to between 5 and 10 feet, so a tree could easily block the passage. Downstream near a footbridge the riverbed deepens and widens a bit. This entire section is dangerous because of fallen trees. One to two hundred yards below the footbridge come some rapids in a slight left turn. Most turns here do have rapids. Roaring Brook enters from the right. Downstream there is a short drop over a rock dam that is easy to run. Another small stream enters at the right; the river then turns left with a chute. The best route is on the right or in the center; an eddy awaits on the right below.

To this point rocks have been mostly bread basket-sized. You should soon spot a slightly larger rock in midstream. Past this rock and the next left turn, a wall of small rocks seems to block the river's passage. This wall is the first of three sets of rapids in close succession. In low or medium water, the best path for the first is on the extreme left, moving diagonally right, down a 1- to 2-foot drop (some 2.5 miles from the covered bridge). This drop may be complicated by more rocks in the path, depending on water level. In medium or high water, other passages will open on the right. Five yards below is another 1- to 2-foot drop over a ledge, best taken either in the right center or on the extreme left if you spot this route quickly enough. Rocks in the center divide these channels. Immediately below, the river narrows, turns right, then quickly left with standing waves in each turn. This three-rapids stretch is the most technically difficult one of the run. Whether to scout depends on the group's strength. These rapids are no more than Class 3, Class 2 in low water.

The portage around the dam at Stewartville is best on the left, although it is mean on either side. Below the dam, a long rapids flows (Class 2 in low water, 3 in high) by a lumber mill on the right bank. From here to West Leyden, the Green shows mostly Class 2 water and the river leaves the road for approximately one mile. Just before the takeout at the bridge in West Leyden, there's a small rock dam that you'll pass over easily.

Green River, Vermont may be reached by following Route 5 to Guilford. Turn west at the center of town and follow the main road until it forks. Take the right road (now dirt) and continue to the covered bridge at Green River. Just before it enters the noise and bustle of this large cosmopolitan town, the road forks. The right fork leads upriver.

The gage is located in Franklin County, 0.5 miles upstream from the bridge at Colrain on the right bank. Look for the gaging station hut beside the road (about 10 feet tall and 5 feet square in cross-section). Then go downhill on foot to the river. The gage is about 20 yards upstream. It is fixed to a rock, faces downstream, and is hard to spot.

Green River
(VT-MA)

WEST LEYDEN TO COVERED BRIDGE
Trip B

Distance (miles)	Average Drop (feet/mile)	Maximum Drop (feet/mile)	Difficulty	Scenery
5.6	31	40	2	Good

TOO LOW	LOW	MEDIUM	HIGH	TOO HIGH	Gage Location	Shuttle (miles)
3.9					East Colrain	6.0

The trip from West Leyden to the covered bridge by the Greenfield water control dam is similar to the trip on the upper Green. However, the lower section is neither so difficult, nor so pretty. There are no special difficulties, and no rapids need be scouted. If the water is up, the run is continuous in that there's little slack water. In low water, the canoeing will be boring to those looking for challenging sport. In all, this is a good trip for open boats. The water control dam is just upstream from the covered bridge and can be recognized by the pool above it. The dam is much too high to run.

For the gage location, see the previous description.

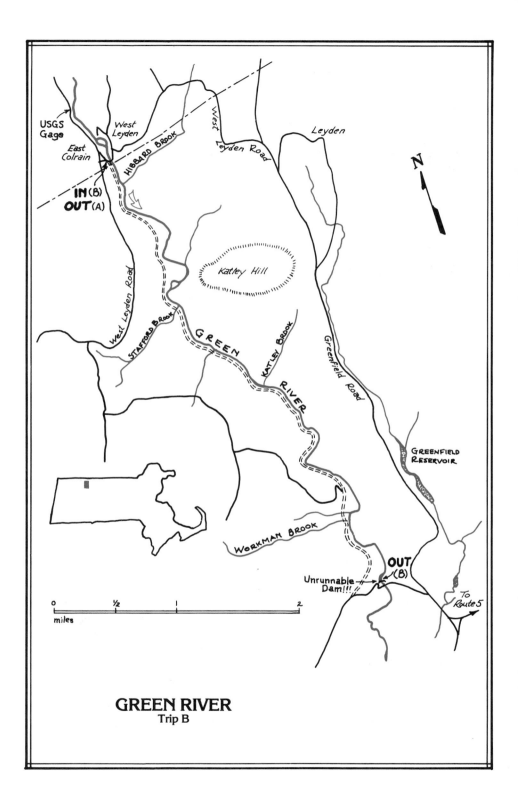

USGS
Gage

West
Leyden

East
Colrain

IN (B)
OUT (A)

West Leyden Road

HIBBARD BROOK

West Leyden Road

Leyden

Katley Hill

STAFFORD BROOK

G R E E N

R I V E R

KATLEY BROOK

Greenfield Road

GREENFIELD
RESERVOIR

WORKMAN BROOK

OUT
(B)

Unrunnable
Dam!!!

To
Route 5

N

0   ½   1   2
—————————————————
miles

GREEN RIVER
Trip B

Hudson River
(NY)

INDIAN RIVER TO ROUTE 28

Distance (miles)	Average Drop (feet/mile)	Maximum Drop (feet/mile)	Difficulty	Scenery
12.5	29	80	4	Excellent

TOO LOW	LOW	MEDIUM	HIGH	TOO HIGH	Gage Location	Shuttle (miles)
2.8		5.5C	6.3C	7.0C	North Creek	14

The Hudson River Gorge is certainly one of the finest runs in the East. Utterly majestic scenery, complete isolation from the daily routine, and challenging rapids all make this trip unforgettable. This part of the Hudson is one of the largest rivers described in this guide, and that means a great deal of powerful H_2O. Once you've started the Hudson trip, it is very difficult to pull out. The course wanders through a river valley that is miles from the nearest road, so you're on your own with any problems that occur. A lost or broken paddle means a long walk — especially tough if there's still snow on the ground. The Hudson's rapids have heavy waves, big holes, and some rocks, depending on the water level. Pools are usually at the end of rapids, but rescues can be difficult due to the river's width and current. Rapids are usually discrete, although in several places they blend together to form extended rapids. The Hudson is a difficult and a demanding run. It is also magnificent.

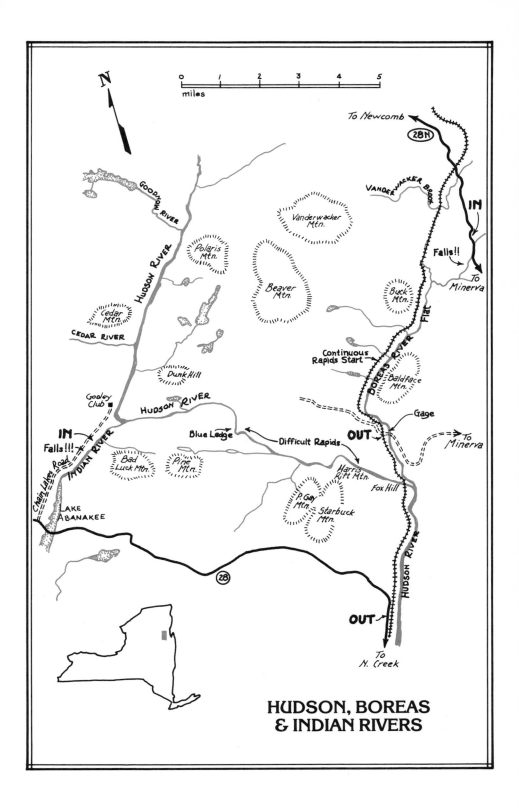

HUDSON, BOREAS
& INDIAN RIVERS

A trip in the "gorge" section usually starts on the Indian River, which joins with the Hudson. To launch on the Indian, turn off Route 28 on to Chain Lakes Road (Class 2, dirt), and go for about 2.8 miles. Along the way you should pass a beach area, a dam at the end of the lake there, and eventually a waterfall. Pick a good spot to turn around and change into wetsuits.

The Indian River is fairly wide itself, with a good distribution of rocks and a discharge that depends on the level of Lake Abanakee above. It is Class 3 under most conditions, a good warmup for the more difficult water to come. After 1 to 1.5 miles, you reach the Hudson. The Hudson is twice or three times as wide as the Indian. A scattering of large boulders and some easy standing-wave rapids greet the paddler. Downstream, note that the sidelines are either solid rock-face or steep, tree-lined slopes. Where there is rock, it would be absolutely impossible to beach a canoe or to effect an exit. Low water reveals gnarled waterline deformities that are superb examples of the water's erosive power. As with some modern-day sculptures, it is best not to stray too close: there are numerous undercut rocks that will pass water, but not a boat or a boater. After some more heavy Class 2 water, a small stream enters from the left and the pace slackens as the riverbed widens. Large rocks start to appear in a slight right turn. In a wide right turn, the pace moves up to a respectable Class 3 as the river falls over small ledges and around rocks. At the end of these rapids, Blue Ledge looms high overhead. Rising precipitously from water level, this bare rock face towers 200 feet upward. Blue Ledge is the stoic sentinel that guards the entrance to more difficult rapids ahead. It's time to empty your boat, rest, and make a quick trip into the woods.

From Blue Ledge to about four miles downriver, the Hudson is a panorama of difficult water. At lower levels, most rapids are technical and demand maneuvering around rocks and holes in fast, heavy water. Rapids usually last for less than a hundred yards, although two continue for .25 miles or more. Most rapids have a pool or some quieter water before the next drop. You may end up swimming, however, to reach the pool. As the water level rises, forget about the rocks. Worry about waves, souse holes, and vicious cross-currents.

Below Blue Ledge, the Hudson bends left into a standing-wave rapids, then a pool. The river turns again after the pool, into perhaps the heaviest drop on the trip, plunging downward among standing waves and souse holes, with large boulders on the sides and smaller ones in the middle. The crosscurrents are turbulent and powerful. The whole thing deserves a look if you haven't seen it previously. It is rated a Class 4 at a gage of 5.0. There is a pool below, a left turn into an easier Class 3 rapids, another pool, then another Class 4 rapids that lasts 25 yards. Again, crosscurrents and turbulence deserve respect. After a right turn comes a long, continuous Class 3-4 stretch that lasts about .25 miles. In medium water, this one is easy for competent boaters, though rocks, holes, and hydraulics gnaw constantly at the boat. Following calm water at the end is an easier Class 3 drop, a right turn, then more rocks and turbulence.

Downstream, the Hudson turns left where a stream enters from the right; then comes some short, good rapids, another pool, a left turn, and then an easy Class 3 rapids.

Shortly after a right turn is the Soup Strainer. One of the trickier rapids on the trip, it has a large boulder in the center followed by a boiling eddy. The right and right center are strewn with large rocks and abrupt drops of 2 to 3 feet. The left is clearer, but turbulent. If you choose right, start out very far right; then thread a "crooked needle" for 25 yards. If you get turned around, back down the rest of the way. Try to stay clear of the boiling eddy in the center because if it catches you it could mash you against the upstream rock as flavoring for the soup. After a few more turns, a right bend starts the longest rapids, which last for about a half mile. At a gage of 5.0, these are Class 3 to 3+ over rocks and hydraulics. The pool at the end marks the last of the major rapids, although there are still some intermediates to punctuate the otherwise calm water that follows.

Below a railroad bridge, the only structure to cross this part of the Hudson, the Boreas River enters from the left. A bit downstream, past where the railroad comes close on the right, the Hudson turns left. On the outside of this turn, extending from the right side, there is a 2- to 3-foot drop, followed by a tough hydraulic. Unless you want to tangle with this tiger, pass on the extreme left. Riding the hydraulic here is like being a surfer on Big Kahuna. Water from here to the take-out, about 3 miles downstream, is relatively slow-flowing (at a gage of 5.0), and paddling is a chore. Past a factory on your right, State Route 28 comes close to the river, a good spot to leave cars for the shuttle. Farther down is the section where a yearly slalom and downriver race takes place.

At a gage of 5.0, this section is no place for a dual-paddled open boat, although a very competent solo paddler could make it without any carries, taking on much water and risking a lost boat. As the gage edges toward 6 and 7, the Hudson changes from a friendly bear to an angry grizzly which clobbers trees in his path. At these higher levels, some people will find real sport, but most people will find huge waves, powerful currents, few rocks, cavernous souse holes, and many opportunities to break the world's record for holding their breath while upside down in a boat. At high levels, the onslaught of water seems to contain enough energy to create a local seismic event.

At a gage reading of around 6, the rapids blend together so that you really cannot tell one from another. The whole run is one of unremitting difficulty: it should not be undertaken by the inexperienced. *Long* swims are possible at these higher levels, and you will undoubtedly wish that you had stayed home with a good book if you get separated from your boat. Also, *if the dam at the start of Indian River is releasing water, the run down the Indian to the Hudson can be as difficult as the Hudson itself.* Be sure to investigate this before you run, because commercial raft trips are starting on the Hudson, and water is released for them. In addition, if the weather is cold, this trip can be downright dangerous for those who don't do everything just right. The Hudson is one of the few rivers where the author has seen *pieces* of aluminum canoes.

The U.S.G.S. gage is in Warren County, on the left bank, 125 feet upstream from the bridge on Highway 28 in North Creek, New York (about 26 miles downstream from Indian Lake). The gage is in the Telemark system. You can obtain the latest reading by calling the special weather service at the Albany airport (518-869-7891). After the recorded message ends, someone will answer whom you can ask for the reading and whether it is going up or down. You can also obtain readings from the Black River Regulating District at 518-465-3491 in Albany.

If you would like to try another river in the area that is also of Class 4 difficulty, the Boreas is a good choice.

Mad River
(NH)

WATERVILLE VALLEY TO GOOSE HOLLOW

Distance (miles)	Average Drop (feet/mile)	Maximum Drop (feet/mile)	Difficulty	Scenery
8.5	85	100+	3-4	Excellent

TOO LOW	LOW	MEDIUM	HIGH	TOO HIGH	Gage Location	Shuttle (miles)
	1.2C	2.5C			6 Mile Bridge	8.5

The Mad is well worth the wait. Beautiful scenery, challenging rapids, and clear water all characterize this fine white water stream. However, there is hardly enough watershed to keep the river filled for more than a few short weeks in mid-spring. But, oh, those weeks! Then the Mad presents one of the nicest, most continuous sets of frothing water in the entire state. Once the water is down, however, boaters just have to wait it out through a long summer, fall, and winter to get another shot at the Mad.

Start your trip on the Mad in Waterville Valley where the road crosses the river. To get there, turn left off Route 49 at a Mount Tecumseh Ski Area sign. There is also a gas station near this turn. The turnoff is almost 10 miles from Campton Pond Dam. At the start, the river is rather small and relatively straightforward. The current, however, is moving at a very healthy pace, since the average gradient is 85 feet/mile, an incredibly large figure. First-time Mad paddlers will find the current decidedly pushy.

Shortly after the start comes a big boulder on the left with a hole right beside it; then, farther downstream, the river turns right, and moves away from the road. There are plenty of rocks, hydraulics, and drops to keep you busy in this loop, but nothing really spectacular. The Mad, then returning to the road, bends sharply right and drops over Ho Hum Rapids. HH Rapids is short and intense. It forces the paddler to make quick maneuvers in a tight channel and a fast current while pounding down a stairstep descent. Ho Hum is Class 4 in medium levels, and still tough in low water. It is the most difficult rapids encountered so far on the trip, and it is indicative of what follows. A roadside turnoff shortly downstream from Ho Hum makes scouting by car easy. Route 49 is on the left.

What follows Ho Hum Rapids is more of the same. If each subsequent set of rapids had a name, this description would resemble a dictionary. No detailed description is given here. Suffice it to say that there are lengths of calmer water, but they are the exception. If the water is at a decent level, boaters will be very busy reading what lies ahead, then trying to act upon it. Several spots of continuously difficult rapids must be taken as they come; they're mostly away from the road, so you can't scout by car. Abrupt stairstep drops, large rocks, and a serpentine course make it necessary to run several rapids without really knowing where the best (in some cases the only) route lies. Planning ahead is a real asset here. If in doubt, get out and walk a bit. The run's continuity and fast pace are exhilarating if all is going well, dismaying if someone needs to be rescued, especially if that someone is you. At medium levels, this trip is rated a strong Class 4. It is difficult without being overwhelming. It is sport, not survival. It is fun. It is intoxicating. It is beautiful.

The approach to Goose Hollow is easy canoeing. The river still has a steep gradient, but the channel is wide and the rocks are relatively small. You can take out at any of the bridges. A larger, modern bridge is being built to replace several smaller bridges here. The construction is changing the river-bed, but probably not enough to matter.

If you would like a longer journey, Campton Pond Dam is another 2.5 miles downriver. The river slows as it approaches the Pond, though. Below the dam is a steep-sided, gorge-like stretch that is quite playful if you want to carry down the slopes. This stretch has been the scene of slalom races. It frequently has water when the upper section does not.

There is a hand-painted gage on a rock on the left bank, just upstream from Six Mile Bridge (so named because it is six miles from Waterville Valley). You can read the gage by looking over the bridge's edge. If you can't see the top of the gage (3 feet), you should think twice about running. There is also a gage measuring the water level at Campton Pond. This gage is in the Telemark system, so the Corps of Engineers can get readings. The gage is not well correlated with the Six Mile Gage, although it is estimated that a reading over 11.0 is needed for a good closed boat run.

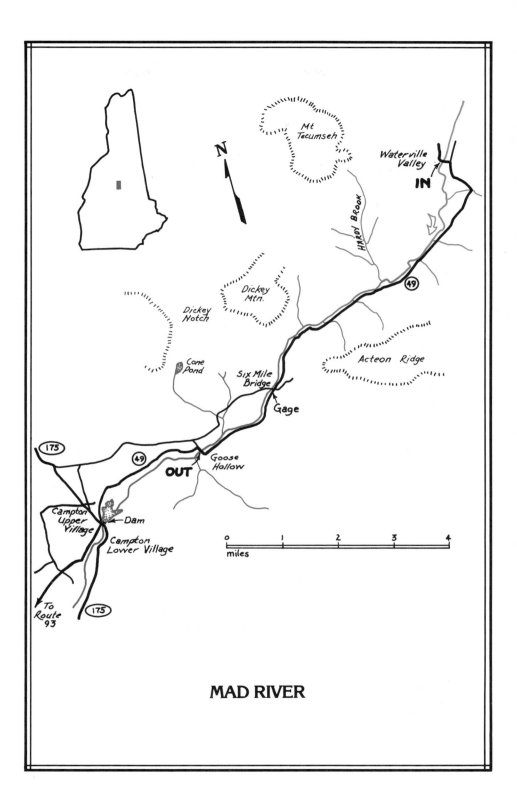

N

Mt Tecumseh

Waterville Valley

IN

HARDY BROOK

49

Dickey Mtn.

Dickey Notch

Acteon Ridge

Cone Pond

Six Mile Bridge

Gage

175

49

Goose Hollow

OUT

Campton Upper Village

Dam

Campton Lower Village

To Route 93

175

0   1   2   3   4
miles

MAD RIVER

Mascoma River (NH)

MASCOMA LAKE TO LEBANON

Distance (miles)	Average Drop (feet/mile)	Maximum Drop (feet/mile)	Difficulty	Scenery
4.0	38	80	2-3	Fair

TOO LOW	LOW	MEDIUM	HIGH	TOO HIGH	Gage Location	Shuttle (miles)
	2.3C	2.3			Route 4A	3.5

The Mascoma River is formed at the outflow of Mascoma Lake, near the junction of Routes 4 and 4A. The lake acts not only as a water source, but also as a flow regulator for the Mascoma, keeping it flowing in relatively dry weather and checking the effect of heavy rains. The river is medium-sized, with numerous fallen trees that narrow the channels. The scenery is only fair — there are many signs of civilization. Except for Excelsior Rapids, the difficulty is rated Class 2-3. Excelsior, which ends the trip, is Class 3 or 4 according to water level. Excelsior is just upstream from a four-foot dam. This height may not be impressive, but the hydraulic at the bottom is. Avoid it. A slalom race is frequently held around Excelsior Rapids.

To start, put in at the Mascoma Lake outlet where there is a little roadside pulloff, reached after driving several hundred yards on a side road off

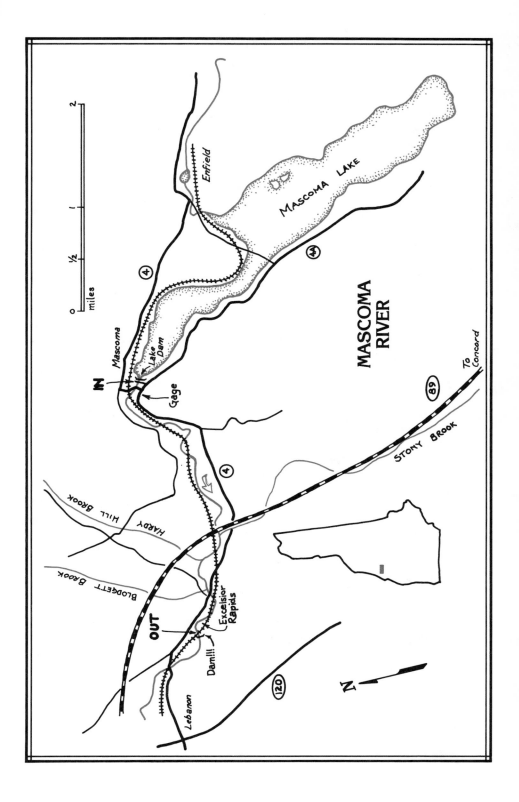

MASCOMA
RIVER

MASCOMA LAKE

Enfield

miles
0 ½ 1 2

Mascoma

IN

Lake Dam

Gage

④

④

④

Stony Brook

To Concord

89

Hardy Hill Brook

Blodgett Brook

OUT

Excelsior Rapids

Dam!!!

Lebanon

120

N

Route 4A. The river starts out rather wide, but it narrows down quickly. Just below the remains of an old railroad bridge, there is a gage on the left. Between the start and the Route 4 bridge are a few standing-wave rapids. From the Route 4 bridge to the Route 89 bridge, rapids are short Class 2-3 (at a gage of 2.3) with spaces of calmer water. There is really nothing here that merits scouting, except perhaps for fallen trees. After the Route 89 bridge, the Mascoma follows the Interstate (Class 1-2).

The Mascoma bears left, away from Route 89, then eventually right again, to begin a more interesting section of river (less than a mile long). After the right turn comes a standing wave rapids under a small stone bridge. There are eddies on either side, and the route is easy to discern. At the right time of year, the boater may see slalom gates starting at this spot. In the right turn below, the river starts a short, tight S-turn with a good hydraulic on the left in the right-turn part, and a few rocks and some good standing waves in the left-turn part. What then follows is one to two hundred yards of Class 3 water, taking you up to a railroad bridge in a right turn. This bridge marks the beginning of Excelsior Rapids.

Excelsior Rapids is harder than anything met on the trip so far, rated Class 4 at medium water levels and Class 3 at low levels. It is about 100 yards long and runs reasonably straight through most of its course, ending in a left turn. The right bank is steep, but there is a footpath on the left bank. Excelsior has everything — rocks, holes, and standing waves that punch back. The usual run starts in a slot on the right or right center, traverses to the left halfway down, and finishes on the left. Several mean hydraulics halfway down on the right are good cause for a traverse. When you sight another railroad bridge after the final left turn, exit from the river: the next right turn immediately below holds the four-foot dam with the powerful hydraulic. Cocky racers whose roll failed them have been known to watch their boats being sacrificed to the river gods there.

The usual take-out is on the left, just before the dam. A short walk across the railroad bridge brings you to a lumber mill parking lot, and from there it's a short walk to the main road. Also it's possible to carry your boat upstream along the railroad tracks for 100 yards and run Excelsior again. Cars with low-slung mufflers had best be careful when they cross the railroad tracks at the take-out parking lot.

From here, the Mascoma meanders lazily down toward Lebanon, New Hampshire, where there is at least one dam. This stretch is not particularly recommended, although there is an outstanding example of some Class 6 rapids that slice their way through town.

The gage is on the left bank, 250 feet downstream from a railroad bridge, 1000 feet downstream from the Mascoma Lake outlet.

Millers River
(MA)

SOUTH ROYALSTON TO ATHOL
Trip A

Distance (miles)	Average Drop (feet/mile)	Maximum Drop (feet/mile)	Difficulty	Scenery
7.0	32	60	2-3	Fair

TOO LOW	LOW	MEDIUM	HIGH	TOO HIGH	Gage Location	Shuttle (miles)
4.4	6.0	7.0			South Royalston	

The Millers River has two sections that are commonly canoed. Both offer exciting paddling if the water is up, but, unfortunately, both sections are polluted. The river passes through several towns where it picks up the offal of local industry. The Millers is an easy winner in the river most likely to see fewest swimmers contest. However, take heart: no boats have been known to dissolve while on the river. Also, it is not quite so bad as the Cuyahoga River which runs through Cleveland and which caught fire in the early seventies, burning several bridges in the process. At higher water levels, the water is not so obviously objectionable. A new waste water treatment plant in Irving does improve the lower part, although nobody is bottling and selling water yet.

The Upper Millers (Trip A) run starts in South Royalston, just a mile below the Birch Hill Dam, which completely controls the discharge. The river offers a pleasant trip in medium water with no major difficulties. It should require no scouting. In South Royalston, there are two bridge crossings; the usual starting spot is below the lower one. Those who wish to start above the first bridge will get to run some fast rapids that slice through town — Class 2 at most levels. Just upstream from the lower bridge is a small dam which should be runnable at a gage of 6.0. At a gage reading of 7.0, the dam is still runnable, but there's a bigger hydraulic. Total drop is 1 to 2 feet. Downstream from the second bridge on both banks are the remains of an old factory. The current in this section is strong at a gage of 7.0; a few rocks show at 6.0. Shortly downstream from the starting point, islands divide the river into several channels, all of which are passable but which might possibly be very rocky if the gage isn't 6.0 or above. Class 2 open boaters will probably be challenged by this section. There are also some small (1 to 1.5 feet) drops here.

After some calm water comes an extended rapids that lasts until the railroad bridge in the distance. At a gage reading of 6.0 this section is scratchy, but at 7.0 it has a strong current with many little hydraulics and rock surfaces showing, calling for moderate rock dodging. The approach to the railroad bridge is rocky, with some nice haystacks just before it where the main stream sprints to the right. These waves are suitable for playing and they are probably the heaviest of the trip. There is a big eddy on the left, downstream side of the bridge. At a gage of 7.0, the run from the start to the railroad bridge is Class 2-3.

The pace slackens after the railroad bridge, and the river swings gently right in the distance. A left turn downstream begins a quarter to a half mile of continuous rapids rated Class 3 at 6.0. Many rocks and hydraulics create ample opportunities for play and practice. Smooth water then predominates until the next bridge is past. As the river approaches the railroad on the left, there are standing-wave rapids in the right turn, a pause, then more rapids. Both channels around the next island are usually clear. After that, the going is easy. As the Millers approaches the railroad again — it's on your left — there are a few easy Class 2 rapids, then, before long, some Class 3 rapids of just about equal difficulty with the long rapids already paddled, though they are not as continuous. Rocks and angled hydraulics prevail. The take-out, just above the first dam and bridge in Athol, is a short distance from these rapids.

When you see houses, stay close to the right bank and get out near several dirt piles. A factory (Union Butterfield) is farther downstream on the left bank, below the dam and the bridge. The take-out, fairly long and steep, leads to a road in a residential area. To reach this road by car from Route 2A, cross the bridge on the upstream side of the Union Butterfield factory, turn right (Route 32), continue for a short way, then take the first right (Crescent Street) until it becomes a dirt road where cars may be parked. For a shuttle, it is best to stick to hard-surfaced roads and avoid the shorter Gulf Road which lies north of the river.

The upper Millers run is much more fluid when the gage reads 7.0 rather than 6.0. At 7.0, most rocks are underwater, the current is uninter-

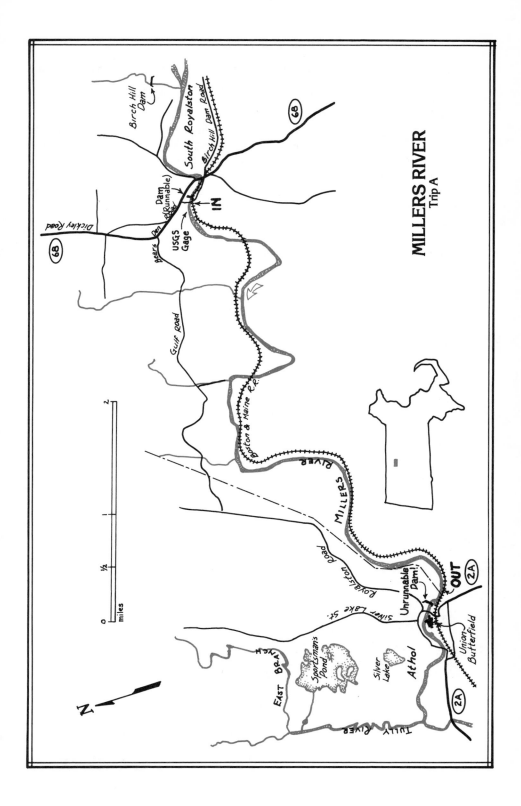

MILLERS RIVER
Trip A

rupted, and the rapids are Class 3 instead of Class 2. People with no experience beyond Class 2 will find the water challenging, and the rescues even more so. Open boats paddled tandem should be no problem at this level (unless they are floating free, or wrapped around a rock).

The gage for this section is located in South Royalston, on the right bank, 500 feet downstream from the second bridge, and 1.7 miles downstream from Birch Hill Dam. The Corps of Engineers may be of use for obtaining information about the discharge.

For another river of similar difficulty, try the Chickley or one of the trips on the Westfield.

Millers River
(MA)

ERVING TO MILLERS FALLS
Trip B

Distance (miles)	Average Drop (feet/mile)	Maximum Drop (feet/mile)	Difficulty	Scenery
6.5	29	64	2-3	Fair

TOO LOW	LOW	MEDIUM	HIGH	TOO HIGH	Gage Location	Shuttle (miles)
2.9		4.4/4.8C			Farley	6.0

     The Lower Millers (Trip *B*), as this section is known, has some of the same drawbacks as the Upper Millers, although it offers better white water sport. The run begins near one paper company, ends near another, and passes by a landfill along the way. How's that for scenery? State Route 2 follows the river for the first half of the trip, and there is a short picturesque valley near the end. Running rapids is mostly riding up and down haystacks with little or no vigorous maneuvering. Some stretches of rapids are long, and one in particular (the Funnel) is very heavy and tricky. Parts of the run are excellent for practicing closed boat surfing techniques or for open boat instruction. This section is also known for holding enough water to be canoeable when other rivers are not. The Irving Paper Company waste water treatment plant has improved water quality, but not to the point of potability.

There are several starting points for this trip. The one farthest upstream is by a bridge near the general offices of the Irving Paper Company. Turn off Route 2 at the sign for these offices, pass under a railroad, and in several hundred feet you will find the bridge in question. From this spot to the first railroad bridge (1 mile) are some easy Class 2 rapids, then smoothly flowing water. This stretch is good to limber up muscles and it takes enough time for your legs to go to sleep. The railroad bridge is an alternative put-in, and the roadside turnoff there is handy for parking and changing — best done on the off-road side of your car lest you cause some commotion on Route 2. A sweeping left turn below holds a long standing-wave rapids, with the largest haystacks toward the end. These waves are good for playing, surfing, and teaching. There's a pause, then another set follows. An island on the left divides the channel downstream, with most of the water rushing to the right side. The road is close at this point, and on a clear day you can see chain-reaction collisions as motorists dream of shooting the rapids themselves. Near here on Route 2 is a sign stating that in the flood of September, 1938 the Millers crested 5.5 feet above the highway. According to the U.S.G.S., this crest corresponds to a discharge of some 29,000 CFS.

Farley Rapids is just below Farley Bridge. The approach is a sharp left turn: the Millers is pushed to the left *below* the bridge in a fast chute that's mostly choppy haystacks and holes, and a bit trickier than it looks (Class 3 at a gage of 4.8). The easy route is on the far right. After passing the Farley landfill on the right bank, you'll see an abandoned gaging station on the right a bit downstream. After several islands comes the first of three closely spaced sets of rapids. The first is the easiest—about 50 yards long and mainly haystacks. A pool follows. The second has somewhat heavier water and stronger crosscurrents than the first. It follows after a right turn, lasts for about 75 yards, and can be run almost anywhere in medium water. Like the first, this rapids is principally haystacks. After a short length of calmer water, the Millers broadens into a shallow, rocky rapids that leads to the Funnel.

The distance from the end of the second drop to the beginning of the Funnel is only several hundred yards, so don't get too close lest the accelerating current invite you in like a smiling cannibal. This drop, known as the Chute, is definitely the hardest of the trip, Class 4 at most levels. It should be scouted. Narrowing quickly, the Millers rather spectacularly drops about five feet in the next 100 yards. In medium and high water, it's a straight shot down a fast, turbulent, violent set of haystacks, souse holes, back-curlers and boiling eddies, and these water formations are constantly changing. Look at it from the right. Look at it from the left. But run it dead center. As is typical with this sort of rapids, rocks line the edges and dot the center route creating abrupt drops, nasty-looking holes, and eddies that could easily disorient a boat. The center has the heaviest water, but it is the clearest run. Three quarters of the way down, there's a sloping 3-foot drop followed by a large wave. The drop is more abrupt on either side than in the center. Twenty yards below is a slight left

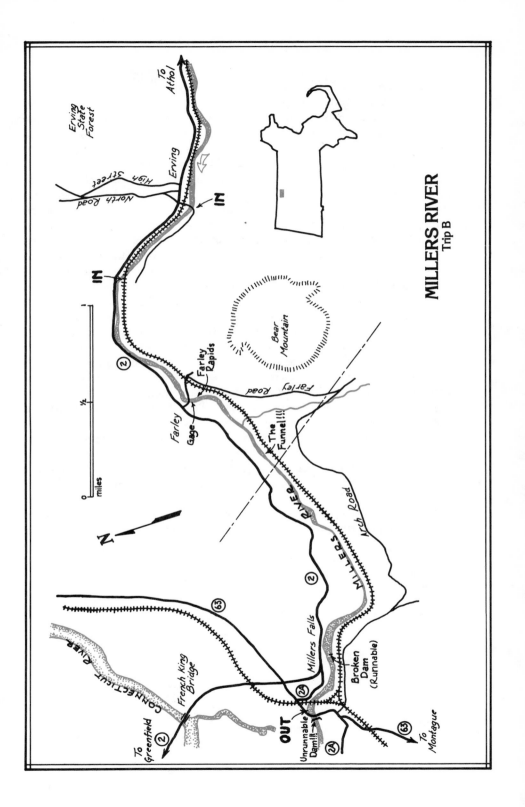

MILLERS RIVER
Trip B

bend and another drop into a large wave. At a gage reading of 4.4, this wave is 3 feet high; it can be taken sideways if you have good balance. At a reading of 4.8, it is more overpowering than the small gage increase would indicate. Hit it on the right, and you can get pushed into a strong eddy by the bank. The Chute takes 10 seconds to run (or swim). Dump out at the top and you'll have a vicious swim, especially at high water, since the outrun is long and powerful. At medium levels, a *good* solo paddler could successfully run the Funnel in an open boat. It is best for a tandem open boat crew to walk. There's a small footpath a quarter of the way up the right bank and a railroad track halfway up the left bank. At a gage reading of 3.0, the Funnel is good Class 3 and can be handled by two in an open boat. At readings of 4.8 and over, even experienced closed boaters should scout, especially if they haven't seen it before. A tree blocking any part of the drop would be real trouble. Below the Funnel a set of powerlines crosses the river, but *do not* attempt to use them to locate this rapids. The lines are too hard to spot: you'll be into the Funnel before you're ready. Downstream from the Funnel are several Class 2-3 rapids and an island.

Route 2 appears high on the right bank as the Millers veers left into Class 3 rapids with haystacks and rocks. Then comes a long stretch of calmer water in an especially pretty section of conifer-lined banks.

The Millers Falls Paper Company dam is broken, and you can run it almost anywhere there's not a rock. Sharpest on the left and in the center, the drop is between 2 and 3 feet. Powerlines overhead mark the spot. You will pass a bridge and in a little while you'll see Millers Falls. Take out on the left, under a railroad bridge where a small road connects with Route 63. Under the next bridge there is a two-stage dam that it is best to avoid unless you enjoy plunging headfirst into a concrete slab.

The gage for this section of the Millers is in Franklin County, on the right bank, 75 feet downstream from the bridge at Farley, 2.4 miles downstream from Erving, and 5.5 miles upstream from the mouth. Go east from the take-out on Route 2, past a small store in Farley (that store *is* Farley), and turn right on the next road — it's a very sharp turn and is close to the store (Bridge St.).

For a river of similar difficulty to the Millers, try the Quaboag or the West Branch of the Westfield, since they are reasonably close.

North River
(VT— MA)

HALIFAX GORGE TO COLRAIN

Distance (miles)	Average Drop (feet/mile)	Maximum Drop (feet/mile)	Difficulty	Scenery
7.0	31	50	2	Good

TOO LOW	LOW	MEDIUM	HIGH	TOO HIGH	Gage Location	Shuttle (miles)
3.3	4.6	5.3			Shattuckville	7.0

The North flows south along State Route 112. It is a pleasant Class 2 run early in the spring season. What rapids there are consist of small waves and easy rock-picking. It is seldom very high and is a good first trip for beginners. It is narrow for its whole length and usually flows slowly, although it has a current most of the way. The scenery, pasture land and wooded areas, is nice but not spectacular. Closed boaters would find the North a real drag.

Put-in is some 2 miles north of the Route 112 bridge which crosses the river near the Vermont-Massachusetts line; here the road goes away from the river, up a steep hill. This spot is just downstream from the Halifax Gorge, which has three unrunnable waterfalls in close succession. To get a good view of Halifax Gorge, continue north from the put-in .20 miles, then park at the turn-off on the west side of the road and follow the 4-H foot trail to the gorge. It is well worth a visit — on foot.

At the put-in, the North is 1 to 2 boat lengths wide and shows as difficult a course as will be seen. One ledge and hydraulic are noteworthy: they will play with beginners. This section peters out shortly into what is more typical of this Class 2 run: occasional small rocks and ledges but mostly just a narrow channel. There aren't even many turns: the North follows a mostly straight course. Just before a small footbridge, several rocks block the outside of a slight left turn. The Route 112 bridge has a grassy island underneath and the North then passes into farmlands, so watch out for bovines, porcines, and equines. This whole section meanders lazily with no special difficulties, while the road plays leapfrog with the river. Take out in Colrain where Route 112 crosses over near a school.

The gage is in Franklin County, on the right bank in Shattuckville, 1.2 miles south of Griswoldville and 1.3 miles upstream from the confluence with the Deerfield. It is just upstream from a small store and bridge in Shattuckville. Just below this gage site is an abrupt drop of 4 to 5 feet over a ledge which a closed boater may consider running in low water, but possible damage to a boat makes it unwise. For lower levels, the appropriate gaging staff is attached to a rock and faces outward, so a bit of rock-hopping is necessary to read it. This gage is in the Telemark system.

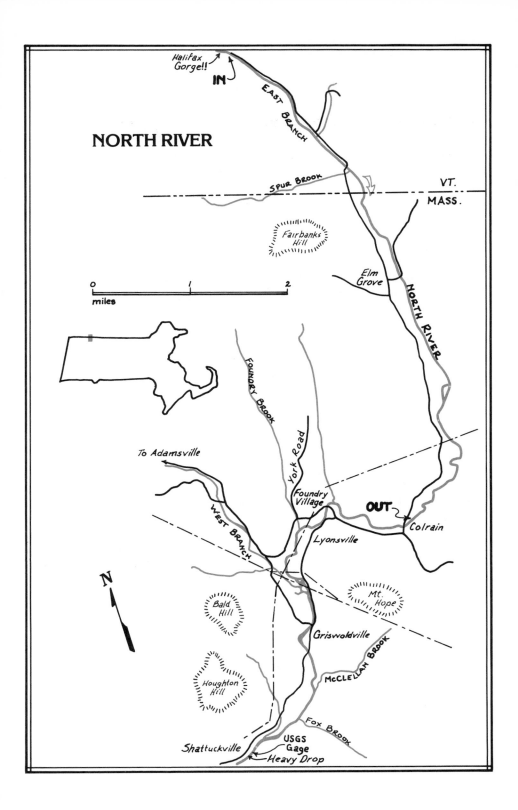

West Branch of the Ompompanoosuc (VT)

STRAFFORD TO RICE MILLS

Distance (miles)	Average Drop (feet/mile)	Maximum Drop (feet/mile)	Difficulty	Scenery	Shuttle (miles)
9	45	60	2-3	Good	9

Any river that has six vowels in its name ought to be something special, but the Omp isn't. Flowing west to east, the Omp is generally mild-mannered — a Clark Kent who never changes into Superman. The river eventually stalls behind a huge Corps of Engineers dam, from which it passes down to the Connecticut River north of White River Junction. There are two branches to the Ompompanoosuc, the East and the West. The East Branch is generally isolated from roads; where it does come close, one can see several unrunnable drops. The West Branch is by far the more practical for canoeing: Class 2-3 at most water levels with a uniform gradient, and it is ideal for open boats.

State Route 132 follows the West Branch closely (Route 113 follows the East Branch), so you can scout the whole trip easily from your car before launching. At low and medium water levels, the rapids are mostly straightforward, and in the Class 2-3 range. Rocks are small to medium, and large portions of the trip are relatively rock-free. In high water the current is fast, and your main problem in an open boat is to avoid standing waves and brush. At high levels, the pace continues with little letup, so rescues will be difficult.

The section from South Strafford to Rice Mills demands more than the one from Strafford to South Strafford; hardy boaters may well choose to ignore

190

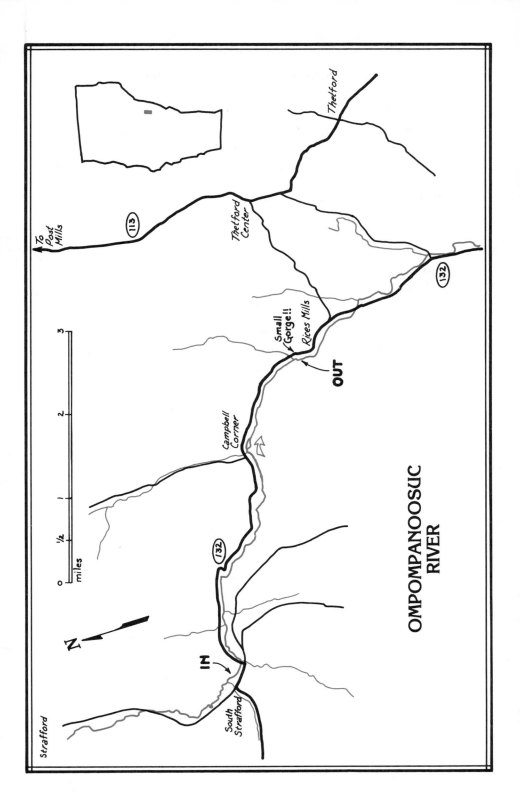

OMPOMPANOOSUC
RIVER

the upper section. The maximum drop of 60 feet per mile is the *average* gradient for this lower stretch, a very healthy figure which deserves respect, especially when the river runs full.

The most taxing rapids occurs at Rice Mills at the end of the trip, where rock walls on both sides form a small, short gorge. The single channel is very narrow, and the current can be fast and rough, depending on the level. You enter this miniature gorge via a right turn over a ledge. The trick to running this gorge is not to catch your bow on any of the protruding rocks. If you do, you'll be in trouble. A good-sized pool follows. This rapids is very close to Route 132, and it is easy to scout or portage.

The West Branch runs down to Union Village Dam, which controls the flow from that point onward. You may contact the Corps to determine the flow discharge as well as the most probable flow into the reservoir. There is no gage that applies for the canoeist on the West Branch.

The East Branch has a higher average drop per mile than the West Branch, but the figure is somewhat deceptive. The high average is due to several large, unrunnable drops; what's in between is mostly easy canoeing. At Thetford Center, there is a 30-foot cascade, and there is an 8- to 10-foot falls farther upstream at Post Mills.

Otter Brook
(NH)

EAST SULLIVAN TO OTTER BROOK PARK

Distance (miles)	Average Drop (feet/mile)	Maximum Drop (feet/mile)	Difficulty	Scenery
3.3	80	100	3-4	Fair

TOO LOW	LOW	MEDIUM	HIGH	TOO HIGH	Gage Location	Shuttle (miles)
	1.5C	1.5	3.0C		East Sullivan	3.3

If the water is up, this trip on Otter Brook presents one of the most delightful closed boat runs in all of New England. The rapids are uninterrupted, with no pools or flat water. There are plenty of rocks, and yet no rapids are extremely complex, though the difficulty can reach Class 4. The water is not usually overpowering or scary, but it is fast and turbulent in a very narrow riverbed — maneuvering is a challenge, and you need to do quite a bit of it. Eddies are plentiful; the competent boater can zip in and out playfully like an otter. Open boats could be used at lower levels, provided that boaters are very careful in several of the harder spots. State Route 9 follows the Otter Brook closely for the entire trip. The rapids are continuous, so only the highlights are presented.

Put in where Otter Brook crosses Route 9 near East Sullivan, New Hampshire. The river is 40 to 60 feet wide here, with gentle rapids (in low or medium water) among a scattering of small rocks. A little way downstream comes the first rapids, a 100-yard collection of Class 3 haystacks, souse holes, and turbulence. This rapids is typical — it puts you in the mood for the rest the run. After a short stretch of the calmer water, the Otter Brook turns a right corner into a long section of continuous rapids, Class 3-4 depending on water level. Waves are up to three feet at medium levels, and large rocks introduce themselves at the very beginning. The road is close to the right bank.

Very shortly after the river leaves the immediate vicinity of the road, there is a right turn and then there's a sloping drop of 3 to 4 feet into standing waves and holes (Class 4 in medium water), followed closely by another right turn and another similar drop. The road reappears and the rapids continue.

Slightly past a brown house on the right bank there is a 2- to 3-foot drop over a ledge in a right turn. Following this are several tight S-curves and some easier drops. This type of rapids continues past an old iron bridge.

The Otter Brook at high water. Would you believe that this paddler is running bow in a C-2? Would you believe that the stern paddler is really buried? Would you believe that both came through this drop in good shape? — Dave Cooper

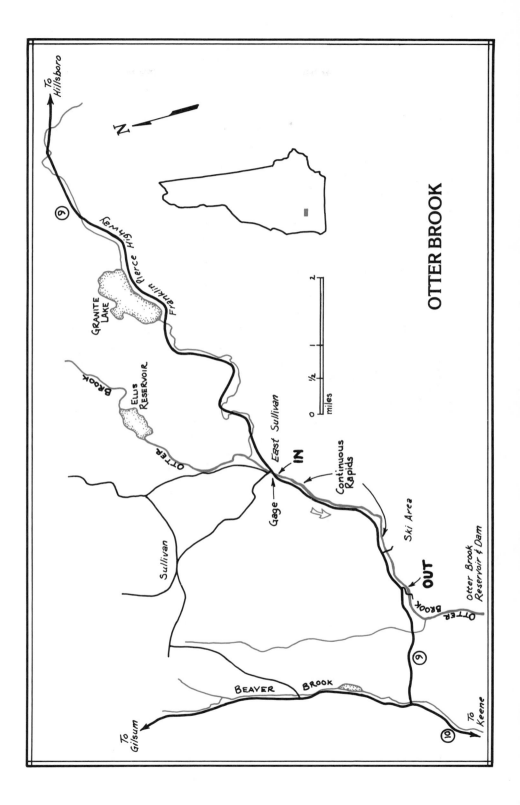

OTTER BROOK

By the time you reach a ski area, the tough rapids seem to have exhausted themselves, and the river becomes a little shallower. From this spot to the take-out, about 0.8 miles, the rapids are less intense (Class 2-3), but the current is still strong and there are plenty of small rocks, though the curves are not so sharp. Take out by the bridge in Otter Brook Park. If the main gate is locked, walk your boat several hundred yards to reach Route 9. If the water is low, you may want to take out at the ski area.

Special mention should be made of the 3-foot gage reading. At this level, the upper part of the run is nearly unbroken Class 4 rapids. Several drops — see the third paragraph above — feel as if the bottom has fallen out, and the elevator is falling free toward an unseen depth. A swimmer will take a pounding and may even resurface occasionally. However, if you are well beyond the learning stage — you no longer lean upstream in a hole — you should find the Otter Brook a truly delightful challenge.

There is a hand-painted gage on the right, upstream side of the center support of the Route 9 bridge in East Sullivan. There is also a Corps of Engineers dam just downstream from the take-out. Informing yourself about the discharge could help you to determine whether the Otter Brook is runnable. It is estimated that 350 CFS will create a medium level for closed boats.

If you would like another nearby trip of similar difficulty, try the Ashuelot or the South Branch of the Ashuelot.

East Branch of the Pemigewasset (NH)

KANCAMAGUS HIGHWAY TO LOON MOUNTAIN
Trip A

Distance (miles)	Average Drop (feet/mile)	Maximum Drop (feet/mile)	Difficulty	Scenery
2.5	72	80	4	Good

TOO LOW	LOW	MEDIUM	HIGH	TOO HIGH	Gage Location	Shuttle (miles)
.2	1.0C	1.5C	2.0C		Kancamagus Bridge	2.5

The East Branch of the Pemi lies on the western edge of the White Mountains, and its valley offers good boating and good hiking. Two characteristics predominate in the Pemi's personality. First, it is a big river — big for the area and big for New England. Second, it has a relatively steep gradient. These two traits combined make a rugged trip for anyone who dares venture forth when there is a respectable amount of water running. Because the Pemi has a large riverbed to fill, it takes a heavy runoff or a recent rain to satisfy its voracious appetite. The Pemi has plenty of rocks, although most are rounded and friendly, or at least as friendly as a rock can get. The Kancamagus Highway follows the riverside for the whole trip, but it can't always be seen. Trip A can be repeated several times in one day; however, if you include Trip B with Trip A, you can have a full day's outing. Open boaters — unless they are skilled or daring — would be better off elsewhere. The rapids combine technical maneuvers and turbulent waves. In medium or high water the turbulence is overpowering and it *cannot* be appreciated by merely looking at the rapids from the road. Because the rapids are uninterrupted, only highlights will be described, so keep in mind that water not specifically mentioned is still difficult.

Travelling west from the Swift River basin, the Kancamagus Highway parallels the Hancock Branch as it descends to meet the Pemi. Many people have looked at this stream, but fallen trees and inadequate amounts of water discourage nearly everybody from trying a trip there. At the put-in on the Pemi there is a convenient roadside turnoff for parking. This parking lot is also at the start of several hiking trails leading into the Pemi Wilderness, one of which travels along an old logging railroad that parallels the upper Pemi to its confluence with the Franconia Branch (3 miles). This section is seldom canoed because there is no access road for canoe-carrying vehicles — a three-mile hike with a boat in your knapsack isn't much fun. This upper section is similar in difficulty to Trip A. The railroad crosses the Franconia Branch and continues up the East Branch in the direction of Crawford Notch. Now back to the Kancamagus Bridge.

At the put-in, the Pemi is some 35 yards wide, and it is loaded with many bread basket-sized rocks. Larger rocks populate the sides. The section of river near the put-in almost always looks scratchy and low, regardless of the water level. If you don't have to do some rock picking here, the rest of the trip is going to be a real challenge. Looking downstream from the start, the Pemi is straight for some 400 to 500 yards, finally turning right in the distance. Other than possibly being fast and choppy, this straight section holds no great difficulties. The right turn has a short, tricky stairstep descent (Class 3-4 at a gage of 1.3) as the Hancock Branch enters from the left.

After Hancock Rapids, you'll be heading toward the 2,680-foot dome-shaped peak of Potash Knob. Just when it seems to be on the immediate right, the Pemi swings left. The channel is turbulent on the right, and hordes of rocks cover the left shore. More of the same follows as the road returns to the river. A short way downstream is a slight right turn with a heavy chute that will bury a bowperson in a C-2 at a gage reading of 1.3 or higher. In lower water, this chute is somewhat more technical.

Below an old gaging station on the right bank, a gravel island divides the river: the main flow is on the right, swinging left, then sharply right at the island's downstream end. There is a rather powerful rapids at this spot. In low water several routes are possible among the large rocks found there. In high or medium water, the rocks are mostly covered, but they are replaced by turbulence. The island itself tends to become submerged at a gage reading of 2.0 or higher. You may want to scout this one if you are unfamiliar with the river. The approach around the island is relatively straightforward if the holes and violent water don't upset you.

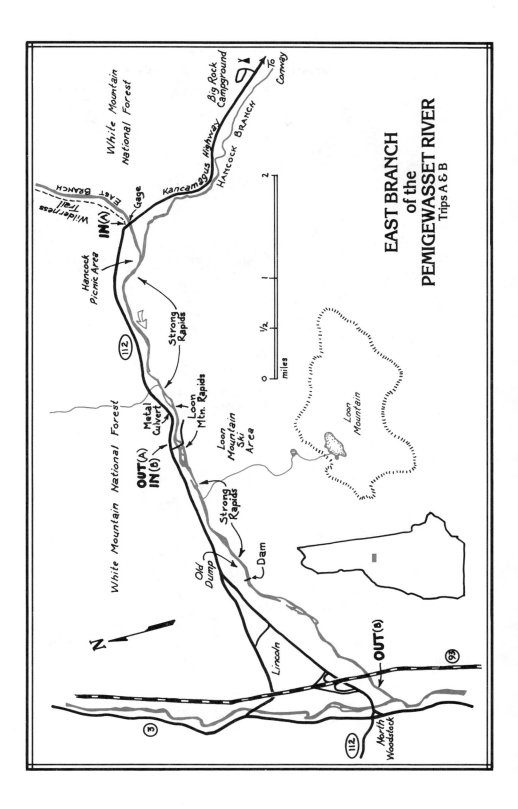

EAST BRANCH
of the
PEMIGEWASSET RIVER
Trips A & B

A characteristic view of the East Branch of the Pemi shortly below the put-in. The water in the photo is low — Lynn Williams

Shortly thereafter comes another gravel island with trees. The right side around the island is narrow and fast with the trickiest rapids on the trip. One-third of the way down are two hydraulics that extend across the width of the channel. Near the end, a collection of three huge boulders blocks the center route. The largest is dead center, and the two smaller ones are slightly upstream to the right and left. The right boulder is underwater at medium levels, so as the paddler comes blasting down this narrow chute he must suddenly make several quick maneuvers through a short, tight slalom course where last-minute decisions can be followed by a swim. The road is very close so you can easily look the whole thing over beforehand. Upstream and downstream from the three boulders you may come upon scattered remains of an old metal culvert, which, although they are sometimes submerged, are a menace at any level. Don't take this right-side channel without scouting it. A trapped tree or other debris in such narrow confines could easily lead to an ugly situation. The left side of the island is another typically Pemi staircase descent. It offers a Class 4 sneak route that avoids the Class 4 route on the right side.

Below, Loon Mountain Rapids starts. This rapids does not look particularly difficult, but it requires *a great deal* of maneuvering around an abundance of rocks, and the paddling is always heavier than it would appear from the road. The take-out is in the middle of Loon Mountain Rapids at the bridge — you can choose left or right for your exit. Loon Mountain Rapids continues beyond the bridge, becoming more malicious along the way, so plan your departure from Trip A carefully (and in advance).

At a gage reading of 2.0 (maybe even 1.5) and above the Pemi deserves special attention: most boaters should exercise discretion, not valor. At these levels the current is savage. The rapids are like barbarians trying to ram the door of a castle, and, if you make a mistake, you'll find the swim more death-defying than that in a moat. At lower levels Loon Mountain Rapids is merely challenging; at these higher levels, it's a malicious labyrinth where the wrong choice of a path leads not just to a dead end, but to a high-speed collision with a boulder as well. The run is continuous Class 4, and the whole trip varies between borderline and definitely Class 5. The paddler must maneuver around numerous rocks and gaping holes in heavy water while trying to stay out of hydraulics that can easily drown a boat. Rescues are nearly impossible. Although eddies exist, they are few and hard to capture. Competent people do boat this section at these levels, but it requires the utmost in skill and performance, and one should be prepared to challenge an angry giant.

There is a hand-painted gage on the right, downstream side of the middle bridge support of the Kancamagus bridge at the put-in. Since the Pemi is a wide river and since the gage is in a particularly wide spot, any change in height means more than it would on a lesser river. Be attentive to small changes in gage height here, especially if the river is rising. There is also a gage on the Pemi at Woodstock which the Corps of Engineers can call. Informing yourself beforehand can save you a three-hour drive only to find an empty river. A rough correlation between the gage at Woodstock and the one on the Kancamagus bridge is given in the following graph.

If you would like a run of similar difficulty to the Pemi (Class 4) try the Swift.

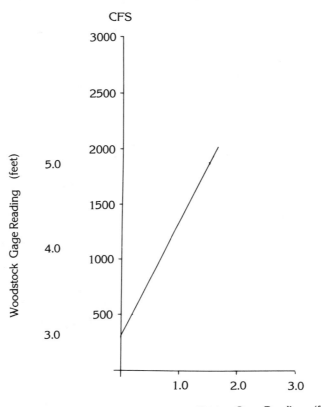

East Branch of the Pemigewasset (NH)

LOON MT. SKI AREA TO ROUTE 93
Trip B

Distance (miles)	Average Drop (feet/mile)	Maximum Drop (feet/mile)	Difficulty	Scenery
3.4	65	100	4	Good

TOO LOW	LOW	MEDIUM	HIGH	TOO HIGH	Gage Location	Shuttle (miles)
	1.0C		2.0C		Kancamagus Bridge	3.4

The Lower Pemi, as this run is known, is similar in all respects to the previous trip. The rapids are just as interesting, maybe even more so; there's a dam either to run or to portage; and, when the water is high, this section will offer a ride you will not soon forget. This trip is paralleled by the Kancamagus Highway, but it is seldom in view. You do get a river's view of the Lincoln town dump and an occasional car in the river. Those skiers just don't know how to drive. The lower Pemi can be run as a continuation of the upper Pemi, or as a separate trip. In either case, it is usually the better of the two runs if the water is low.

Starting at Loon Mountain bridge, the paddler is immediately faced with difficulty, namely the continuation of Loon Mountain Rapids. The area below the bridge was once the site of a dam, but what a recent flood didn't destroy, the bulldozers did. Now the river is relatively wide, housing an immense boulder field. There are two main channels here, and they run close to either bank. The middle tends to be dry except in high water. Both channels are difficult, filled with holes, hydraulics, sharp drops, and lots of rocks. In low water, the left side is a bit easier, but it is also longer. The right side is shorter, but it has more difficult drops. In high water, trying to decide which side is harder is about as difficult as telling which of two maddened attack dogs has the most unpleasant breath. The whole rapids is prone to change, so give this section a good look. Loon Mountain Rapids usually paddles harder than it looks.

Shortly downstream from Loon Mountain, the whole river condenses to pass through a narrow slot. In low water, this spot is a Class 3-4 rapids where exposed rocks force the boater to maneuver in a heavy current. A small change in the gage reading will mean a big difference in the character of this drop.

Next is Decision Rapids. A gravel island in the middle of the river has been broken by the current in at least four places, and you have to choose which route is best, a harder decision in low water than in high. The river generally moves from left to right, and each route has its own set of peculiarities. The whole area is unstable and changeable, so pay your money and take your chances. Scout before running.

The Pemi is relatively calm for a while, then a left turn starts some more action in an extended drop. There are lots of rocks in low water and standing waves in high. After a pool comes more good rapids — relatively long and definitely exciting. Again, much maneuvering is required, so boaters should count on being really busy to avoid trouble spots. At a gage of 0.8, this rapids is rated a good Class 3 or Class 4. In high water consider this stretch a real bruiser.

The old Lincoln town dump is next. Located on the right bank, it signals an upcoming dam. The dam has a rock ledge extending out from the left bank, and the current continues right up to the edge. The total drop is around 4 to 5 feet. It has been run on the extreme left over the rock ledge, but look it over for yourself.

From this spot to Route 93, you will come upon several islands and more rapids similar to what has already been met. The river will narrow and widen several times because of islands; when it constricts, be ready for a roller coaster ride. After the Route 93 bridge and the railroad bridge below, it is a short distance to the take-out on the right bank near a small grocery store (cannot be seen from the river) on the outskirts of Lincoln. To find the take-out from the road, proceed on the Kancamagus Highway past Route 93 until you see the store on the left. There is a small dirt road to the left of the store which leads to the river.

The lower Pemi becomes a real bear in high water and should be given infinite respect. The difficulties are continuous and the entire run approaches, or jumps into, the Class 5 range. Rescues will be very difficult. The trip should be attempted only by strong groups.

There is a hand-painted gage on the middle downstream bridge support of the Kancamagus bridge at the start of Trip A.

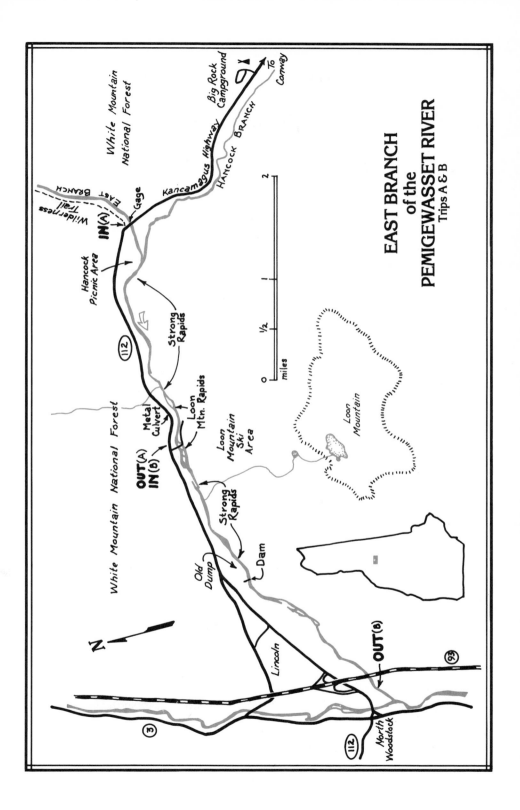

EAST BRANCH
of the
PEMIGEWASSET RIVER
Trips A & B

White Mountain National Forest

Big Rock Campground

Hancock Branch

To Conway

Kancamagus Highway

EAST BRANCH

Wilderness Trail

Gage

IN (A)

Hancock Picnic Area

Strong Rapids

112

Loon Mtn. Rapids

Metal Culvert

OUT (A)
IN (B)

Loon Mountain Ski Area

White Mountain National Forest

Strong Rapids

Old Dump

Dam

Loon Mountain

Lincoln

OUT (B)

93

3

112

North Woodstock

N

2
½
0
miles

Piscataquog River (South Branch) (NH)

NEW BOSTON TO GOFFSTOWN

Distance (miles)	Average Drop (feet/mile)	Maximum Drop (feet/mile)	Difficulty	Scenery
8	13	30	1-2	Good

TOO LOW	LOW	MEDIUM	HIGH	TOO HIGH	Gage Location	Shuttle (miles)
315		318 6.8			Route 13 Grasmere	8

The Piscataquog is a small river in southern New Hampshire which is frequently used as a training ground for beginning white water canoeists. It is usually runnable during mid-spring, and it presents no special difficulties when it is finally freed of ice. The gradient is uniform, turns are gentle, and the hardest rapids come at the start of the trip. Route 13 follows alongside the river for nearly all of the run, so most of the river can be scouted beforehand. The river can be run at almost any level, and water problems are usually due to a lack, rather than to an abundance, thereof. As the Piscataquog approaches Goffstown, the road disappears, along with the current. There is a high dam in Goffstown after the take-out. The dam causes no problems, but paddlers should be aware that it's there.

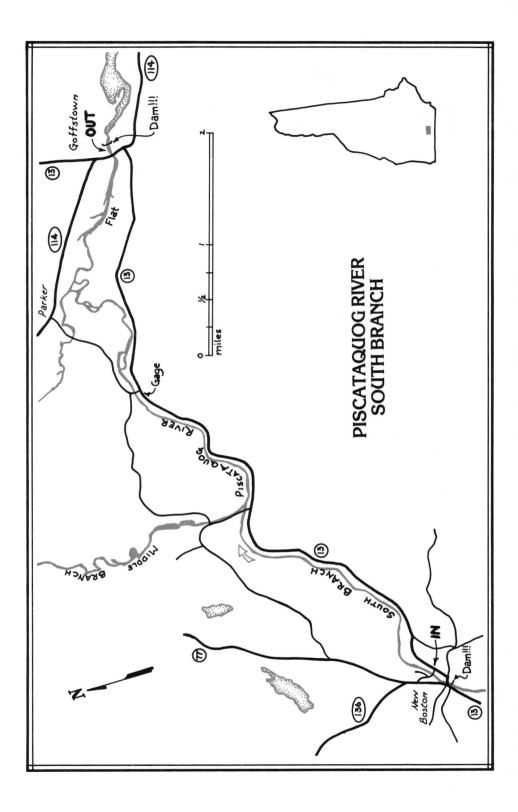

PISCATAQUOG RIVER
SOUTH BRANCH

Start this trip just downstream from New Boston by a metal bridge. On the left bank, set back a bit, is an old railroad station. In the center of town, the river drops over a dam and then waddles its way between abrupt stone walls that make for a tough beginning. At the put-in, the river is about 40 feet wide at medium levels, and maybe half that at lower levels. The current is straightforward, with a few bread basket-sized rocks scattered about the riverbed.

Downstream from the put-in, the Piscataquog turns left, and in this turn there are rapids as hard as you'll see on the whole trip. A few rocks in the center at the start mark the rapids. The current swings left as the river turns, and then comes quickly back right again. The whole rapids take on an S shape. In medium water, standing waves are plentiful and will slop into a tandem paddled open boat if you attack the waves directly. After the last right turn of the S, the Piscataquog continues on its way, and Route 13 loops out of sight for a while.

The Piscataquog then flows past several bridges and several islands, never exceeding Class 2 difficulty. A U.S.G.S. gaging station appears on the right bank just before another bridge. There is no outside staff, but there is a painted gage on the upstream part of the bridge supports. This spot is about 4.3 miles from the start. From this spot onward, there are no remarkable difficulties with the possible exception of downed trees. As the river approaches Goffstown, the current slackens and Route 13 passes from view.

Take out on the left side, just downstream from the bridge in Goffstown, where a small dead-end road parallels the river for a short way. There is a large dam just below this spot.

The main gage for this section is painted on the upstream supports of a bridge 4.3 miles downstream from New Boston on Route 13.

There is also a U.S.G.S. farther downstream. To reach this gage, go west from the junction of Routes 114 and 114A for 1 mile to a flashing yellow light. Turn right toward Grasmere, and proceed to the river. The gage is 300 feet upstream from the bridge there, on the left side.

Quaboag River (MA)

WARREN TO ROUTE 67

Distance (miles)	Average Drop (feet/mile)	Maximum Drop (feet/mile)	Difficulty	Scenery
5.5	31	85	2-3-4	Poor

TOO LOW	LOW	MEDIUM	HIGH	TOO HIGH	Gage Location	Shuttle (miles)
	3.9	4.4 5.5C	5.5	6.0	West Brimfield	6.0

The Quaboag is certainly one of the finer canoeing rivers in the western Massachusetts area. Medium-sized, with rapids of sufficient power to demand great respect, the Quaboag holds its water well and is canoeable in the fall during the rainy season. There are many rapids on the Quaboag, and water level will determine the kind of run to expect. In high water, less maneuvering is involved, although keeping an open boat from swamping in the many standing waves and souse holes is tough. This level offers the most challenge for closed boats. Medium water exposes more rocks and makes several rapids more technical. The rapids are in general spaced by calmer waters, giving the paddler ample time to rest. This trip on the Quaboag flows through several towns, so the scenery isn't great.

Put in at the Lucy Stone Park, which is about .5 miles north of Warren on River Road. Directly in front of the park is an island with a small chute on its left side that can be run by starting above the bridge there. Below the put-in, the water is Class 1-2 until a group of boulders announces the bridge at Warren. Directly below the bridge is a 6-inch drop (low water), starting a standing wave rapids (Class 2). One hundred yards below a railroad bridge is another 6-inch drop, starting some short rapids in a right turn. Another railroad bridge follows. From this point to the first portage around an old dam, the river moves swiftly with intermittent haystacks but no rapids of consequence.

The dam has an outflow on the left through several large tubes, and this area should be avoided at all costs. On the extreme right, two concrete and stone pillars, separated by three feet, pass the remainder of the water. This narrow slot is known as the Mouse Hole; it can be run, although it is very tight and should be looked over first. The drop here is a rather abrupt three feet, ending with a hole and a haystack. Get sideways in the approach, and you'll have real trouble. Go through it upside down in a kayak, and you'll be in a very select club. The outrun is fast with standing waves. At one time there was a stone wall to the left of the Mouse Hole, but it's now falling apart. Depending on the water level, you may even find a route over its remains. Be aware that the Mouse Hole Rapids exists and that it can be very dangerous. Scout it so you can make your own decision. Fifty yards below the Mouse Hole is a small drop under a bridge.

The next major rapids, Trestle, appear beyond a railroad bridge which is in a slight left turn. You should scout Trestle if you haven't seen this rapids previously, especially if the water is high. There are two main routes you can follow. Both are, in some respects, easier at high than at medium levels. For the less thrilling (though still an exciting) route, hug the right bank under the bridge and pass along the concrete foundation, making a sharp right turn into a narrow channel to the right side of an island in the right side of the riverbed. The turn is quite tricky and, if you make it, the rest of the passage around the island is straightforward. If you don't make it, the rest of the way is still straightforward; you just have a harder time of it. As the water level goes down, rocks guarding the approach to this path (and the turn itself) make things more difficult. For the more exciting route, pass either to the right or left of the bridge support and immediately move to a position left of center in the main channel. Follow the flow through standing waves and rocks and, in high water, be careful of a series of holes near the end of the rapids. In low water, these holes are filled with rocks; you can go around either way.

Shortly downstream is a broken dam that's easy to run on the right side. The current angles first right, then left. This route has haystacks at first, then a small island divides the river into rocky Class 2-3 rapids on either side. Farther downstream the next dam appears. Portage on the right. This is not an easy carry, but it is easier than running the 15- to 20-foot drop. Calm water upstream signals the approaching dam. The water below the dam is Class 2, then Class 1, past several bridges and a factory.

Looking downstream at the approach to the Mouse Hole. Notice the current going to the left just in front of the Mouse Hole itself. This is a dangerous spot so scout it — Bruce Arnold

Looking upstream at the exit from the Mouse Hole. There isn't a lot of room even if you run it right — Bruce Arnold

Below a railroad bridge there is a fairly sharp left turn, and then comes a large island with rapids on either side. The left side is slightly easier. The next main obstacle is another broken dam (3 feet high). This dam can be run to the left of two stone columns on the extreme left since it is completely broken there. Going over the dam itself is also possible, but it involves a somewhat greater risk. Downstream, past a green bridge, the river veers left away from the road, passes under another railroad bridge, and enters a 75-yard chute with haystacks that will swamp most doubly paddled open boats at HIGH levels. This rapids is known as Angel's Field. In low water, small rocks stick their heads up, with the greatest density being near the bottom of the drop. In general, they are not a problem. A large pool below is handy for recovery. Before this drop, there is also a sewage treatment plant that sometimes adds to the discharge of the river.

The drop at the end of Devil's Gorge section on the Quaboag — Bruce Arnold

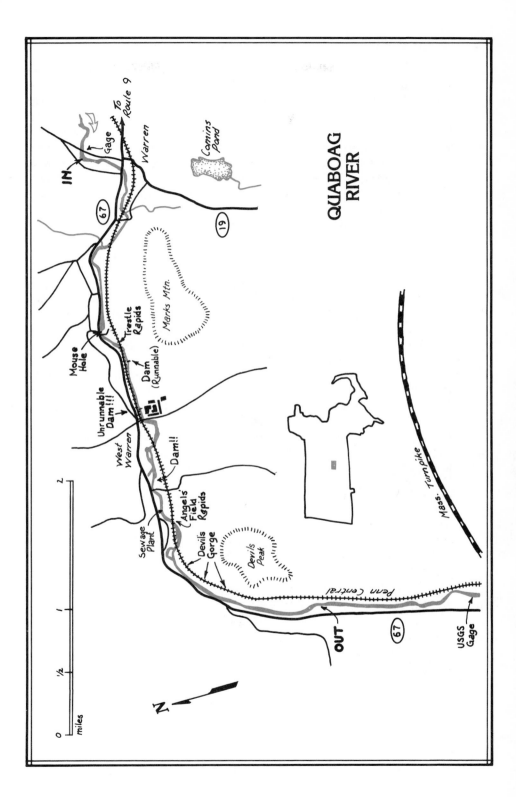

QUABOAG RIVER

Below the pool, the river turns right, passes under yet another railroad bridge, flows around an island, and then enters the first of three good sets of rapids. At higher levels these three stretches merge into one long stretch, with only brief lulls in between. This is the Devil's Gorge section. The first rapids has many small rocks that show up at low levels, choppy waves in medium water, and an abrupt drop (Quaboag Drop) at the end at any level. You can't see this drop from above, and it could be a big surprise if you don't know it's there. It is sharpest in the center, so run it on either extreme. The drop is not so sharp in high water as in low water. In high water it is followed immediately by a 3- to 4-foot haystack. The long approach to this drop can be fast and sassy. A short stretch of quickly flowing but relatively calm water follows, and then the next of the three rapids. This one is about 100 yards long, Class 3 in medium water, and similar to the previous rapids. Another breather, then the last of the trio appears and nothing new is to be encountered.

Because of its length, this stretch is difficult to scout. In addition, it is not at all easy to see from Route 67, which is fairly close. At a gage reading of 5.5, this section is probably too heavy for an open boat paddled tandem. A solo paddled open boat can make it if the paddler is very careful to avoid the larger waves and holes. At this level the entire section is rated Class 4. Take out at one of the several roadside turnoffs that follow.

The West Brimfield gage is in Hampden County, on the right bank, 10 feet upstream from an abandoned highway bridge site, 0.9 miles upstream from Blodgett Mill Brook, and 3.5 miles northeast of Palmer. It can be reached via the last roadside turnoff before the turnpike (about 0.8 miles from the pike). Don't be too surprised if the gage isn't of much help. Some enterprising person has sawed off the bottom portion. A new gage has been painted on the left, upstream side of the Lucy Stone Bridge at the put-in.

For another run similar in difficulty to the Quaboag, try the West Branch of the Westfield or the Millers.

Rapid River
(ME)

MIDDLE DAM TO CEDAR STUMP

Distance (miles)	Average Drop (feet/mile)	Maximum Drop (feet/mile)	Difficulty	Scenery
4.5	40	80	4	Excellent

TOO LOW	LOW	MEDIUM	HIGH	TOO HIGH	Gage Location	Shuttle (miles)
		1200 CFS/C			Middle Dam	4.5

It certainly can be — rapid, that is. Heavy turbulent water, huge holes, and a scattering of rocks for good measure characterize this fine white water river. The Rapid really is special, as a well-known Boston canoeist often states. It starts out from Lake Richardson, drops quickly to Pond in the River, and then plunges again to Lake Umbagog. It is this latter section that houses most of the heavy water. For closed boats this is truly one of the best playgrounds in the Northeast. Normally you wouldn't expect a river that starts and finishes in a lake and passes another on the way to have such fine white water, but there it is. The blue sky, the green vegetation, and the white water make this one of the most aesthetically pleasing trips anywhere.

Because of its isolation in the Maine woods, the Rapid is moderately difficult to canoe. Arrival requires a one-hour motorboat ride across Lake

Umbagog, which can deposit the paddler at a semiprimitive campsite named Cedar Stump, located just below the last set of rapids.

For information and reservations concerning the boat ride and camp-site, try contacting the Dustin Store on the Magalloway River (603-482-3388) near Errol, New Hampshire. If this doesn't work, try Brown Owl Camps at 603-482-3274. The responsibility for these services seems to bounce around a little, so keep trying. It is also possible to paddle the lake, which can be choppy or smooth. From Cedar Stump, it is 4 to 4.5 miles to Middle Dam on Lake Richardson, where most white water trips begin. There is a road (Carry Road) which runs this length, although it is generally well back from the river itself. At one time Lakewood Camps provided a shuttle service for boats up this road, but its status is unclear now. This camp is located on Lake Richardson near Middle Dam. When you call (207-243-2959), be aware it is sometimes difficult to get hold of them. If the shuttle service isn't working, then you are on your own. The best way to hook up with Carry Road from Cedar Stump is to paddle downstream a short way, take the first right, and locate a small path leading back into the woods on the north side of the little cove there. There is also a trail that runs out of the upstream side of Cedar Stump and parallels the river for some distance before joining Carry Road.

Middle Dam is owned and operated by the Union Water Power Company of Lewiston, Maine (207-784-4501). You should contact them about water releases. If contacted several months in advance they will sometimes hold water back to insure a large group will have enough. The dam is a wooden structure in various states of repair, just waiting for Hans Brinker. The water immediately below is very turbulent and the eddy near shore is almost a whirlpool. The easiest put-in is below the dam on the right side, just around a slight right turn in the river. Anyone who attempts to run the dam, even as a stunt, has rocks in his head, or soon will have.

At 1,000 CFS, the drop to Pond in the River is mostly heavy water — Class 3 — over a distance of about .5 miles. The water is a series of moderately turbulent standing waves with a few rocks and a sweeping left turn. At Pond in the River, the current stops, and a 1.5-mile paddle across flat water ensues. The outlet to this lake used to be Lower Dam, but now it is just a hulking skeleton of old memories that can be run just about anywhere (although the passage can be shallow, even at 1,000 CFS). The middle channel has the largest drop, about 1.5 feet. On the right below, there's a nice small notch for resting and swimming. A turnoff from Carry Road leads to Lower Dam, where trips may also be started if you don't want to paddle the lake.

The outflow from Lower Dam is smooth and shallow, leading shortly to an easy Class 2 rapids. For the next mile, till the first of the big drops, the paddler passes several islands and a few summer homes on the right shore. Rapids are mostly Class 2-3, and some can be shallow. Even though the water is relatively easy, a detailed inspection of the water's motions will forewarn the

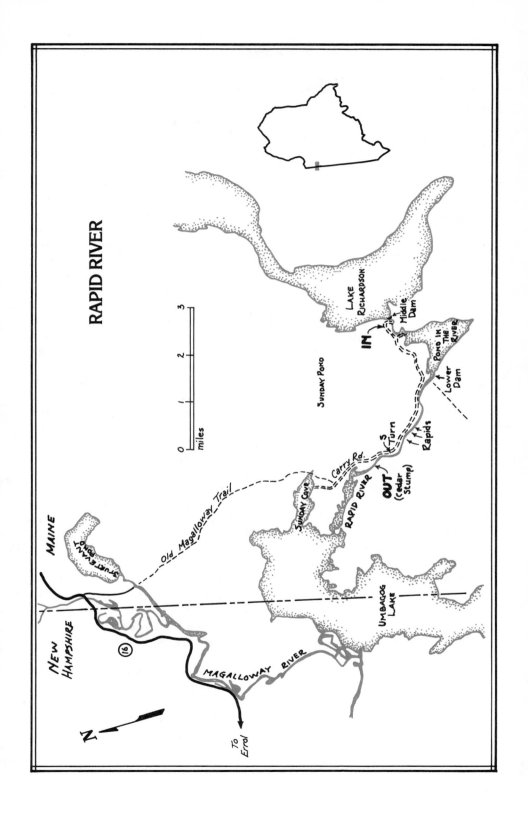

RAPID RIVER

boater of what lies ahead. Enshrouding all creatures within, the Maine woods march right up to the river's edge in one of the few really isolated white water spots in New England.

The first big drop (Class 4) takes place in a fairly straight section and it is easy to recognize. Slanting in a slight angle to the right, this rapids is about 100 yards long and is made up of large, irregular standing waves and souse holes. Rocks are scattered about but generally remain invisible until the last second. The safest run is dead center, where the turbulence is greatest. As with the other rapids here, there are more rocks on the sides, creating abrupt drops, so a boater looking for an easy route will find trouble instead. At the end is a left turn and, in several hundred yards, the next big one. The heaviest rapids on the trip, the second drop starts out with standing waves measuring 4 to 6 feet at 1,000 CFS. Again, they are in the middle but can be safely avoided by canoeing in the left center, around them. Run'em head on and they'll drive you vertical — both ways. The rest of the drop is quite similar to the first — mean. After 100 to 200 yards of quieter water, the next goodie appears. Assuming your eyes have been open so far, this one doesn't display anything different from those above: turbulence, engulfing souse holes, and several lines of rocks crawling out from the right bank halfway down. Negotiating these three rapids is the acrobatic equivalent of a drunk running a bobsled track on roller skates.

Shortly below the third drop is Smooth Ledge. Whether you're swimming or paddling, it is a great time-out place for lunch, sun worshipping, or playing a hydraulic. Smooth Ledge is a large flat rock advancing from the right, ahead of the woods. It extends underwater to midstream, creating a nice hole-curler combination for the more energetic to play in, so bring along your sand pail and shovel. It is usually not big enough for enders but respectable for pop-ups. Dropping on a slant of about two feet with a two-foot wave following, it can be avoided on the left side.

Immediately below Smooth Ledge, one sees more white stuff with the waves flickering upward, licking an ethereal face. This is the beginning of the famous S-turn, which is the most difficult of the course. Slanting downward, the paddler first sees a several-hundred-yard section that's Class 3-4, but a blind right turn hides all hell breaking loose. For approximately .25 miles, this whole rapids is as deceptive as a grandmotherly con artist. Previously, most standing waves were just that — standing waves. Here there's a good possibility that they're pillows hiding rocks, and when you charge through ##$*!!. The water is very fast, with few eddies and much turbulence. It is somewhat shallower than previous rapids mainly because the riverbed is wider. Near the end, before the final flick of the S, the river turns right. Stay center or right here and then left again, falling into a blaze of standing waves, foam, and spray at the end. Following is more smooth, fast water forming a large, flowing, pool-like expanse that empties to the right into an easy Class 4 rapids. This is the first of perhaps six sets of rapids spaced between short stretches of calmer water. This series extends .5 to 1 mile, and the rapids are rated Class 4 at 1,000 CFS, probably a little lower with less water. Although not so turbulent as the first three drops, the ever-present, boat-devouring rocks and pushy crosscurrents still make them difficult. The rapids end rather abruptly, and Cedar Stump is then but a few minutes' paddle away.

There is no gage on the Rapid. The flow is regulated entirely by Middle Dam.

Sacandaga River (NY)

Although it is not located in New England proper, many clubs in the Northeast enjoy the Sacandaga. Situated in the Adirondack Mountains, north of Amsterdam, New York along Route 30, the Sacandaga offers three branches that differ as much from each other in character as do three sisters. It is usually run over a weekend so that the river connoisseur can fully savor each and every trip. The West Branch is by far the largest and contains the heaviest water. The Middle Branch is smallest and provides the shortest trip. The East Branch is the quickest and hardest if the water is high. The usual campsite is the Sacandaga Public Campsite, situated right on the river where the outflow from Algonquin Lake and the West Branch meet, about 2 miles below the town of Wells. Although a Sacandaga weekend is usually rated Class 4, actual difficulty will be determined by the water level. Even if the water is high, there are sections that are less than Class 4 in difficulty. The real disadvantage of the Sacandaga, however, is the time and effort it takes to get there, especially if you happen to live near the Atlantic.

West Branch of the Sacandaga (NY)

WHITEHOUSE TO CAMPGROUND
Trip A

Distance (miles)	Average Drop (feet/mile)	Maximum Drop (feet/mile)	Difficulty	Scenery
8.5	36	70	2-3+	Good

TOO LOW	LOW	MEDIUM	HIGH	TOO HIGH	Gage Location	Shuttle (miles)
	3.3	5.0C	5.0S	5.0	Hope	12
	3.1	5.5C	5.5S	5.5	Griffin	

The longest and heaviest water is to be found on this branch. Large standing waves try to overpower any canoe, and even the smaller waves pack tremendous force that the canoeist can't really appreciate until he has had a rather brusque introduction. Superimposed upon this is a current which is amazingly fast-paced for the river's size, especially in high water, so your reaction time is cut down slightly. Fortunately very little maneuvering is required unless you attempt to avoid the bigger waves or the souse holes that populate the run.

To start the trip on the West Branch, proceed from the campsite north on Route 30 toward Wells and turn left just before the town, near the outlet of Lake Algonquin. Continue until the road forks. Follow the left fork and go about 8 miles to Whitehouse over a road that could be snow-blocked, depending on the time of the year. If you can't reach Whitehouse, put in by the area where the road leaves the river for the last time near Jimmy Creek. Probably the heaviest rapids on the trip is immediately downstream. This rapids is large, with powerful standing waves all across the river; it can be run almost anywhere, although an open boat would not stand much chance in high water. If you run this rapids successfully, the rest should be easy: remaining rapids are similar, although not so heavy, with an occasional souse hole thrown in for good measure. You can avoid most big waves if you want to, and the run is otherwise straightforward.

When the West Branch joins the outflow from Lake Algonquin, the campsite is on the immediate left bank, so a quick ferry is all you need for an afternoon's nap or another run. In low water, this entire section can probably be handled by an open boat. Difficulty will be in the Class 2-3 range.

There are two gages on the Sacandaga. One is on the East Branch near Griffin, in Hamilton County, on the left bank 300 feet upstream from the highway bridge at Griffin, 2 miles downstream from Georgia Creek, and 3 miles upstream from the mouth. It is just above the spectacular East Branch Gorge. The other gage is on the main river near Hope, on the left bank 1.5 miles downstream from the West Branch and 4.5 miles upstream from Hope. An observer relays daily Hope readings to the Hudson River-Black River Regulating District in Albany. You can contact them at 518-465-3491.

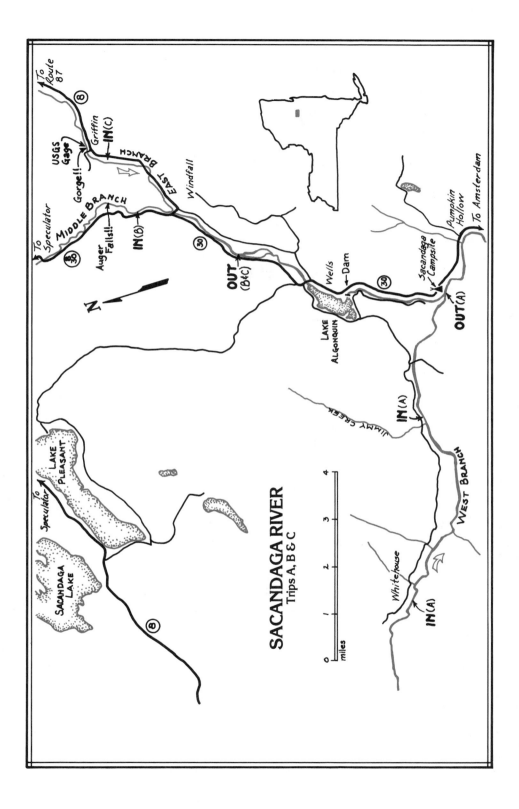

SACANDAGA RIVER
Trips A, B & C

Middle Branch of the Sacandaga (NY)

ROUTE 8 TO ROUTE 30
Trip B

Distance (miles)	Average Drop (feet/mile)	Maximum Drop (feet/mile)	Difficulty	Scenery
2.7	43	50	3	Good

TOO LOW	LOW	MEDIUM	HIGH	TOO HIGH	Gage Location	Shuttle (miles)
3.1		6.4C			Griffin	3
3.3		5.5C			Hope	

To reach the Middle Branch go north through Wells on Route 30 and then onto Route 8. A short way past the Route 8 bridge, a small road angles off to the right, leading to the river. The river here is perhaps 50 to 60 feet wide, flowing swiftly among a wide scattering of rocks. The rapids start shortly downstream. In high water they are very closely spaced, choppy standing waves. The pace is not quite so fast as it is on the East Branch, but it is fast nonetheless. Just before the approach to the Route 8 bridge, a large island divides the river. Narrow channels exist on both sides. The excitement increases as the current accelerates.

After it joins the East Branch, the river widens considerably and speed diminishes, although there are still standing waves if the water is high. One

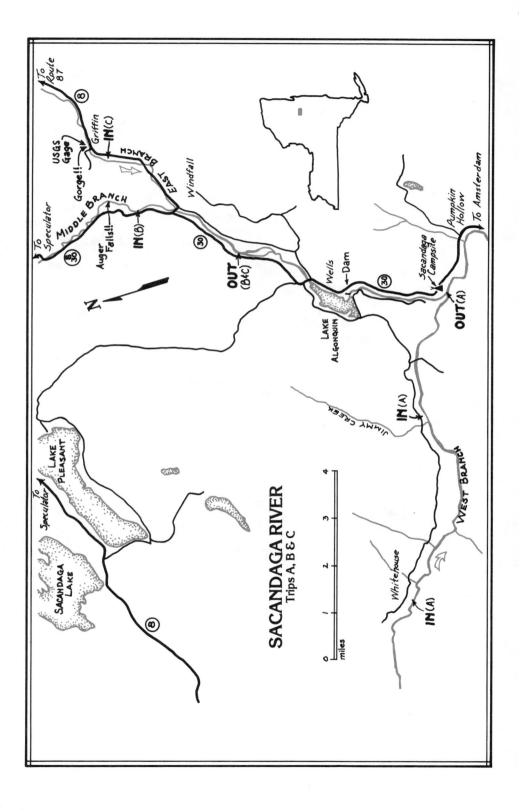

SACANDAGA RIVER
Trips A, B & C

could, then, enjoyably continue, taking out where Route 30 approaches the right bank.

It is also possible to make a longer run on the Middle Branch by starting higher up, since Route 8 stays close for many miles. If you decide to do this be prepared for at least one dam and one waterfall. Both can be seen from the road if you are looking.

Photo - Lynn Williams

East Branch of the Sacandaga (NY)

ROUTE 8 TO ROUTE 30
Trip C

Distance (miles)	Average Drop (feet/mile)	Maximum Drop (feet/mile)	Difficulty	Scenery
2.0	55	80	3-4	Good

TOO LOW	LOW	MEDIUM	HIGH	TOO HIGH	Gage Location	Shuttle (miles)
3.1		5.5C	6.4C		Griffin	2
3.3		5.0C	5.5C		Hope	

The East Branch may be reached by proceeding north from Wells on Route 30 to Route 8, turning right, crossing the bridge over the combined Middle and East Branch, and continuing on until you find a desirable starting point. This section is fairly short and can easily be run several times in a day. Route 8 parallels the river for the entire length, which is handy for the initial scouting. Put in above the rapids section, where the river is just a small stream meandering through a marshy terrain. The current is swift but nothing like it is lower down, where an entrance would be somewhat less graceful. The river then widens a bit and begins to pick up speed like a runaway truck gathering momentum. Standing waves are short and choppy at first, graduating to big and choppy. The speed with which the boat is stampeded downward is impressive. If the water is high, eddies are scarce.

About halfway down, a group of rocks gathered in the center and right requires a bit of negotiation to avoid. In high water they form a nice, abrupt drop in the center, with an upsetting hydraulic immediately following. Lying in a left turn, pass them on the left for the easiest route. In lower water, more rocks are exposed, demanding a much trickier maneuver. These rapids are easy to spot from the road — there is a little turnoff there — look them over. The East Branch then continues on in its typical hurricane fashion, with waves and souse holes, to the Route 8 bridge where the worst, or best, is over.

In high water, the whole run is one where, if you can keep your bow pointed generally downstream, you should do OK. Easier said than done. Fierce waves and strong crosscurrents constantly divert the boat. Any maneuvering must be planned and executed well in advance, lest the extremely swift current rush the paddler headlong into trouble. In lower water, rocks naturally appear and the river becomes more technical. Any separation of a paddler from his boat in the upper stretches may result in a long swim, since other boaters probably could not give much aid. In the spring the Sacandaga can rise a foot quite easily in one day.

There are two gages on the Sacandaga. One is on the East Branch near Griffin, in Hamilton County, on the left bank 300 feet upstream from the highway bridge at Griffin, 2 miles downstream from Georgia Creek, and 3 miles upstream from the mouth. It's just above the spectacular East Branch Gorge. Lasting only a hundred yards, the walls of the gorge rise precipitously and the water plunges violently down a series of cascades laden with huge boulders. It is certainly worth a visit. The other gage is on the main river near Hope, on the left bank, 1.5 miles downstream from the West Branch, and 4.5 miles upstream from Hope. Neither gage is in the Telemark system but an observer relays Hope gage readings to the Hudson River-Black River Regulating District in Albany. The latest reading may be obtained by calling them at 518-465-3491.

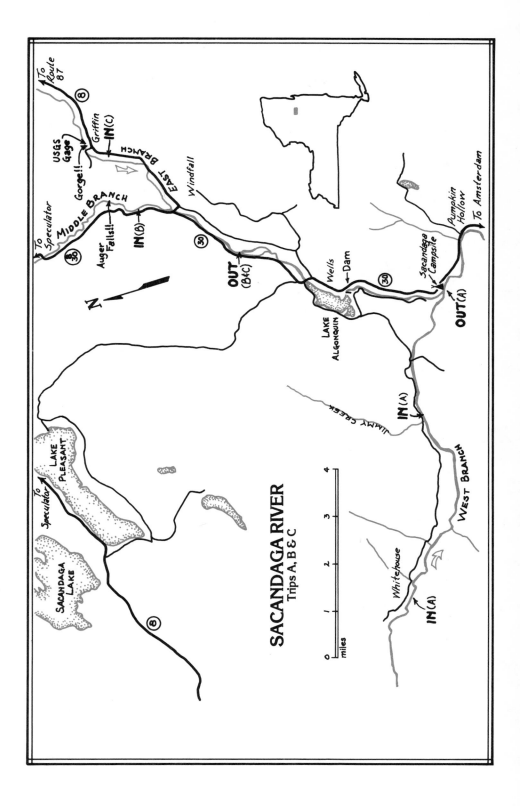

SACANDAGA RIVER
Trips A, B & C

Saco River (NH)

CRAWFORD NOTCH TO BARTLETT

Distance (miles)	Average Drop (feet/mile)	Maximum Drop (feet/mile)	Difficulty	Scenery
6.5	40	60	3-4	Excellent

TOO LOW	LOW	MEDIUM	HIGH	TOO HIGH	Gage Location	Shuttle (miles)
0.5	1.0C			4.0C	Bartlett	6.5

The watershed valley of the upper Saco is one of the most impressive in all New England. Driving west on Route 302 near the river, the traveler is ringed on all sides by tall peaks, as if some mythological giant rested his crown on the earth, encircling the entire valley. Giants in their own right, these peaks possess such non-mythological names as Hancock, Jackson, Clay, Jefferson, and that titan from Massachusetts, Webster. On a clear spring or summer day, the majesty of this view is breathtaking, which accounts for visits by many landscape artists.

The Saco River itself is born out of Saco Lake, in the middle of Crawford Notch. It flows generally southeast to Conway, then on to the Atlantic. Canoeists, however, have scant opportunity to enjoy the scenery, because the rapids offer such good white water sport. For closed boaters the Saco is at its best when the water is roaring, which usually happens when the other White

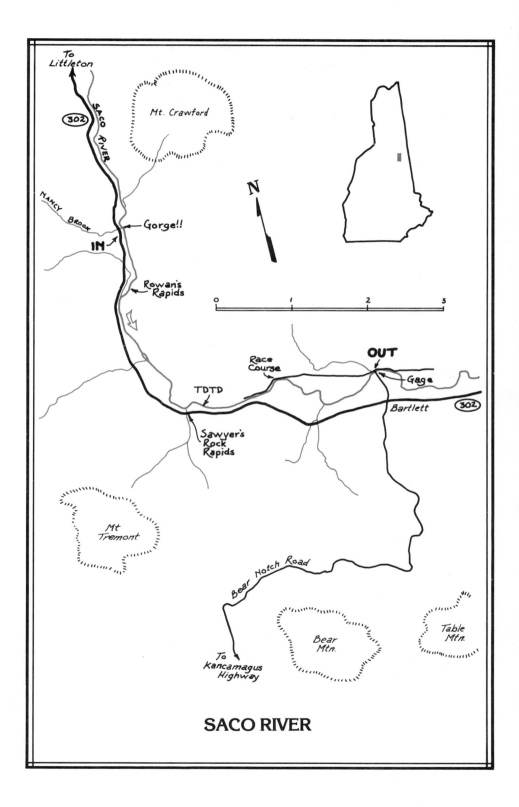

SACO RIVER

Mountain rivers are too high for safe boating. At these levels, the run is unremittingly fast, and rescues can be really tough. The rapids are choppy standing waves and countless souse holes. These conditions prevail when the gage approaches 3 and beyond. At lower levels rocks show themselves, and the run becomes more technical and more interesting to open boaters. At a gage reading of 1.0 the run is rated Class 2-3, and it offers numerous opportunities to pin your boat, closed or open. Regardless of level, be aware that spring thaws can raise the river level dramatically within hours. An enjoyable Class 2-3 trip can change into a Class 4 challenge if you don't stay alert to what's happening.

To start a trip on the upper Saco, put in south of Notchland, just downriver of the Saco Gorge where a small stream (Nancy Brook) passes under Route 302. You can see the last portion of the gorge by looking down this stream. The gorge itself has been run by closed boaters in low water, but this is not really recommended. As the water level rises, so do your chances of getting hurt. Most of the passage is narrower than a boat's length; the walls are solid rock, and there are several sharp 2- to 3-foot drops scattered around. It is well worth a visit — on foot.

At the put-in, the "trail" from the highway to the river is down a steep bank to a railroad track and then down an even steeper bank to the river — a solid Class 4 put-in at any water level. At this point the Saco is about 75 feet wide and flowing fast from the drop through the gorge. The first rapids, which commences immediately, is a very long run through haystacks in a fairly straight section of river. In high water these waves are quite pushy and turbulent. This section terminates in a sharp right turn where the current tries to force the paddler over a small drop on the outside. Eddies rest upstream on the left, and on either side downstream from the drop.

Directly below this turn is a brief, smooth-flowing area, followed by another drop, heaviest in the middle and easiest on the left. Arriving shortly is another long rapids that is broken by a calmer spot squeezed in the middle. These two sections comprise Rowan's Rapids. The first stage is easier: it consists of haystacks and rocks. The second stage is heavier, but similar. You should scout it in high water if you are uncertain or have a weak trip. A cluster of rocks in the middle, halfway down, causes sudden drops into holes and tricky crosscurrents. In high water these rocks are partially covered and deceptive. At the bottom is a left turn. You can see Route 302 here, high on the right shore. Lower down the hill there is an older unused road. The entire section just described is continuous, rated Class 4 in high water. At a gage of 1.0, it is Class 3 with lots of rocks. Some people would even go so far as to call it annoying at this lower level.

From here to the trip's end, the Saco displays more difficult sections, but they are generally spaced with calmer stretches. The run will be rocky and technical if the level is low, or full of standing waves if the water is high.

The first leg of the put-in for the Saco trip. The second part is just as steep — Debbie Arnold

Below, in a slight left turn, there is a 50-yard chute with shifty haystacks, and, farther on, two more sets of similar rapids. As the Saco approaches a high dirt bank on the left and turns right underneath, there is still another chute. Depending on water level, this is Class 2-4, and the most obstacle-free path is in the center. A railroad bridge is next as the river turns left. On the right, the Saco drops about 3 feet (high water only) over a rock into a very ugly-looking hole which is just waiting to devour anything or anyone that might pop in for dinner. Following this is Sawyer's Rock Rapids.*

In a left turn, a large rock ledge extends from the right bank, creating an abrupt drop that tapers down gradually as it moves out to the center of the riverbed. This is Sawyer's Rock. It is run most easily on the left. Hydraulics and haystacks await below. One hydraulic in particular is rather tricky. In high water it can stop a boat or even pull it back in if you're not careful. A big eddy sits just below Sawyer's Rock. This area is also usually full of curious onlookers and crying kids. Sawyer's Rock Rapids is easy to see from Route 302.

Several hundred yards downstream from Sawyer's Rock is Tweedle Dum-Tweedle Dee. Named for two large boulders 30 to 40 yards apart in the left and left center, this rapids comes on the outside of a right turn with an island on the inside. The right side of the island is a trivial route. On the left, the main channel speeds up and forms a series of powerful haystacks (up to 3 feet in high water) and holes. Dee is upstream and not much of an obstacle while

* Sawyer's Rock was named after Benjamin Sawyer, who in 1771, after pushing a horse through Crawford Notch for the first time, performed the ecological equivalent of throwing a beer bottle against the rock.

Dum is downstream and closer to the center with its upstream pillow pouring into a boat-eating hole. Ten yards below Dum is another sizeable standing wave-plus-hole combination to catch those who have relaxed a bit. Run TDTD in the right center. It is rated Class 3 when the gage reads 1.0.

Just in front of the next railroad bridge is another drop creating some huge — up to 4 feet in high water — waves in the right center. To avoid them, stay on the left. Below, the river takes a sharp right where the last heavy water awaits — standing waves, Class 3 or 3+. This is sometimes the site for a slalom. From here to the take-out, the Saco meanders 1.5 miles or so, mostly Class 2, between low-lying gravel banks. Tucked into this Class 2 stretch is a little Class 3 surprise that will trap the unwary. This rapids has lots of rocks and requires some maneuvering. The channel itself is S shaped. Take out near the Bartlett bridge by a dirt parking lot on the left side.

The gage is a hand-painted one, on the right, downstream support of the Bartlett bridge. From time to time this gage gets covered up by rocks. There is another gage on the downstream side of the center bridge support, and the two seem to have a rough correlation.

If you are looking for other rivers with Class 4 difficulty, try the Ellis, Swift, or Pemi. If you want Class 2-3, then try the Ammonoosuc.

Jeremy – Salmon Rivers (CT)

OLD ROUTE 2 TO COMSTOCK BRIDGE

Distance (miles)	Average Drop (feet/mile)	Maximum Drop (feet/mile)	Difficulty	Scenery
5.8	27	50	2	Good

TOO LOW	LOW	MEDIUM	HIGH	TOO HIGH	Gage Location	Shuttle (miles)
	1.4	3.0			Comstock Bridge	6

When winter skies blow away, and the ice is transformed to a more pliable medium, the traditional season's opener is the Salmon. Located southeast of Hartford, the Salmon is usually run early in March and can provide a course for training beginners at the start of the spring season. Usually entered via either the Blackledge or the Jeremy River, the Salmon and its smaller associates can provide an easy Class 2-3 run in medium and high water, or a somewhat uninteresting trip in low. In any case, a rusty paddler has ample opportunity to recall skills left behind in the fall without being severely challenged. Since the Jeremy usually has more water and is slightly more interesting than the Blackledge, it is the generally preferred route and will be discussed at length here.

The trip on the Jeremy, and then the Salmon, may be started where old Route 2 crosses the river, south of Marborough and north of North Westchester. Here the Jeremy varies between one and three boat lengths wide and flows gently around meanders arched with trees leaning toward the river as if saluting a passing procession. The river continues in this way until it passes under the new Route 2 bridge, about .33 miles below the start. Here the river turns 90° to the right, and Meadow Brook adds its water from the left as the Jeremy widens and deepens with the going still easy. Downstream, note several large rocks in the middle and left as the Jeremy approaches the highway on the right bank. Then, the river turns left, and the rapids begin. They are fairly easy and consist of rock-picking for several meanders, with a short pause and then another rapids in a slight left turn. A row of dilapidated houses high on the right bank watches over shallow rapids that last about 50 yards. This whole section is rated 2-2+. Next an iron bridge is passed; the current slows to a stop; the river turns right and a dam in North Westchester must be portaged, best done on the left. Shortly below the dam come more easy rapids; then the river enters a short hemlock-shrouded valley housing some rock-dodging rapids. After a right turn, there is a 1- to 1.5-foot ledge extending from the left bank to midstream. It can be avoided on the right. If the water is high enough, the ledge can be run almost anywhere since it slopes gently downward. This section ends in a river-wide ripple followed by a quiet pool; then come easy rapids that drop into a conifer-lined pool. Except for the possibility of fallen trees, there are no substantial difficulties from here to the Blackledge.

The confluence of the Blackledge from the north and the Jeremy from the east forms the Salmon, which is wider than both of its tributaries. The Salmon begins gently, and the first rapids occurs several hundred yards downstream. Probably the most technically difficult rapids normally run, this set has rocks on either side of its entrance, so choose an initial center route. After passing these rocks, ease a bit to the left. This rapids ends with several rock portals on the right, where there is also a nice eddy. A competent paddler should have no trouble here. (If the water is up he may want to play.) The Salmon then continues to flow easily with rocks interspaced, passing under a bridge where there is a small waterfall on the right.The take-out spot is along Bull Hill Road (River Road) by a brick fireplace. If the venturesome want a little more excitement, they can continue for another couple of hundred yards and run an old broken dam.

The dam is upstream from a high dirt cliff on the left and downstream from a small stream that enters right. There is a broad, shallow approach over a rock shelf, then the dam itself — three distinct ledges which can be run in various spots depending on the water level. The first and third ledges are fairly uniform in their 1-foot drop but the second drops more abruptly (1.5 to 2 feet in several places). In low water, the easiest run is a chute on the extreme right. In any case, stop and look this one over. Continue on and take out by a covered bridge, just upstream from the Route 16 crossing.

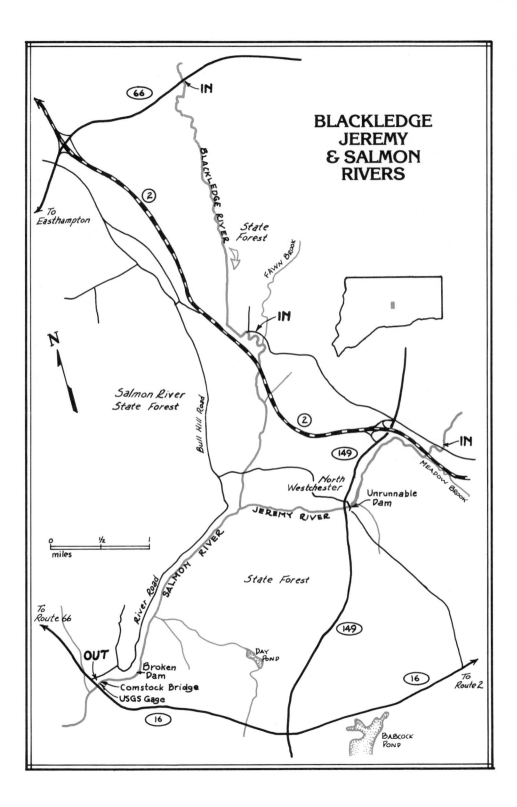

BLACKLEDGE
JEREMY
& SALMON
RIVERS

The gage is located on the right bank, attached to the downstream side of Old Comstock Covered Bridge near Route 16.

Other rivers in southern Connecticut in the Class 2-3 category are the Blackledge, Ten Mile, and the Shepaug.

Boating on the Salmon River in the early spring. The snow cover is not unusual for this time of the year, and it makes for a cold sport — Bruce Arnold

Sandy Brook
(CT)

SANDY BROOK ROAD TO ROUTE 8 BRIDGE

Distance (miles)	Average Drop (feet/mile)	Maximum Drop (feet/mile)	Difficulty	Scenery
3.0	80	100	3-4	Good

TOO LOW	LOW	MEDIUM	HIGH	TOO HIGH	Gage Location	Shuttle (miles)
2.1	2.6	3.3C		3.3	Route 8 Bridge	3

Sandy Brook is a small but extremely interesting stream that runs into the Still River just above Riverton, Connecticut. For its size, it has more rocks than even the upper Farmington and it's just as technical, maybe more so. A continuous current, several narrow, intricate chutes, and a plethora of rocks force the canoeist to make split-second decisions time after time. In many cases, a hesitation or wrong choice will result in a very unpleasant situation. Although it's only about three miles, the trip will seem much longer, especially in low water. Sandy Brook Road follows alongside almost all the way, so a weary paddler can take out any time after exhaustion overwhelms the rock gardener. Again, as with the upper Farmington, water level is critical.

A trip may be started anywhere along the stream, but the suggested put-in is near a bridge some three miles upriver from the gage. At the put-in, the Sandy is very narrow and flows swiftly, with the first rocky rapids found shortly downstream where a smaller stream enters from the right. About a third of a mile below the start, in a left turn, is a 3- to 4-foot drop that must, in lower water, be approached on the extreme right of the chute. Just in the entrance pull sharp to the left to avoid a rock and then . . . down. The drop is tight and should be looked over. Portage if in doubt: it would be foolish to ruin the trip here. It can be seen from the road. Below is a pool, then another drop on the left. One-third of a mile farther down, a mass of low boulders in the center splits the channel in a right turn. The right course is clearest, but ends in a drop of one to two feet over a ledge; it is too dry to run in low water. The chute on the left is intricate — little rocks make a nuisance of themselves. It starts off with an abrupt 1.5- to 2-foot drop and ends in a fast, clear channel. Look this one over also.

The Sandy continues on with an assortment of rocks and tight courses that make it necessary for the boater to continually move his boat. In one left turn, there is what appears to be a broken rock dam with a substantial rock garden below it where the passage is quite intricate. This rapids is about 50 yards long, and proper water level is important. After the left turn, start out left of center and then do your best. These rapids lead to the first of two bridges, about a quarter of a mile apart. Between the two bridges are more small drops requiring a little maneuvering. The first drop comes just after the first right turn. Below the second bridge are more rocks.

About 2.2. miles from the start, just above a lumber yard, and in a right turn with a steep left bank, is another tricky chute requiring a good approach and several crisp turns. Start on the extreme left and proceed to the center. Near the bottom, a large rock divides the river: go around it. A pool awaits below, and, in the distance, so do more rapids in a left turn. The rest of the way to the Route 8 bridge is much easier but still quick. Either take out there or continue down the Class 2 section to the Still and eventually the Farmington.

As with the upper Farmington from Otis to New Boston, this river becomes hellish in high water. Although there are shallow quiet spots, the rapids tend to blend together. The rocks are so close that even the water has to plan ahead for the best way to navigate the course. Precise boat placement is absolutely necessary and you have to know well in advance where you're going. The pace is very fast and requires constant work with few or no breathers. In some spots you can even add a touch of good old-fashioned turbulence. Unfortunately, the Sandy has a very short season.

The gage on the Sandy is on the left bank, attached to the downstream side of the Route 8 bridge that crosses the stream at Robertsville.

If you want another Class 4 run in the general area of the Sandy, try the Farmington above New Boston.

SANDY BROOK

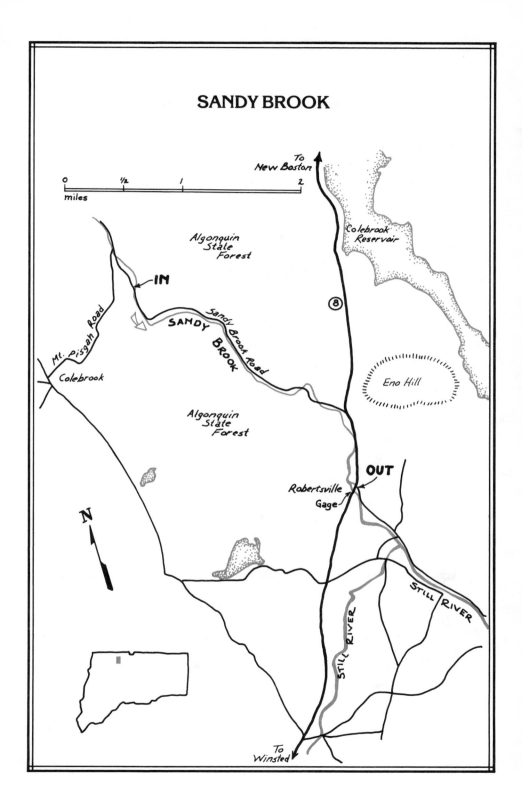

Saxtons River (VT)

GRAFTON TO SAXTONS RIVER

Distance (miles)	Average Drop (feet/mile)	Maximum Drop (feet/mile)	Difficulty	Scenery
7.5	46	60	2-3	Fair

TOO LOW	LOW	MEDIUM	HIGH	TOO HIGH	Gage Location	Shuttle (miles)
4.2		5.2			Saxtons River	7.5

If there is enough water, the Saxtons can offer one of the sportiest runs in Vermont. The upper part is nearly uninterrupted Class 3, yet no rapids are sufficiently difficult to require scouting, though several do call for fast maneuvering. The lower section is mostly Class 2, but punctuated by several trickier drops. In all, the Saxtons is ideal for an open boat, paddled doubly or solo. It is also excellent for learning technical maneuvering in a closed boat. Even closed boaters will get a thrill in medium or high water, since haystacks replace the rocks and the current speeds up to make a strong Class 3 trip. The valley is not particularly attractive. Many bridges cross the river. A road runs alongside most of the way.

Start your trip in Grafton where Bridge Number One crosses the North Branch. Here the Saxtons is barely two boat lengths wide and moves only at a moderate pace. The rapids start immediately below the bridge — standing

242

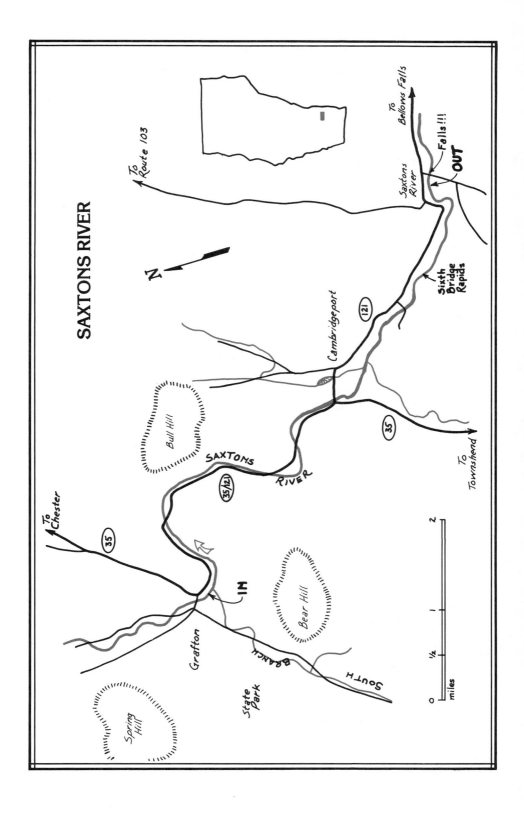

SAXTONS RIVER

waves running about small rocks. Shortly, the South Branch adds its waters from the right and the main river is formed. A covered bridge, typical of Vermont's countryside, is visible just upstream on the South Branch. Below, the rapids continue, picking up a bit. A particularly nice set comes up after a left turn, as the river approaches some houses on the left. A big rock squats in the center here. After more good rapids you'll pass the second bridge. When the Saxtons appears to run into the right bank, the water turns sharply left and drops over a 6-inch ledge, then continues in Class 2-3 fashion.

Below the next bridge (number three) are some small ledges, then there's a really sporty rapids in a slight right turn with a 1-foot drop at the end. A big rock in the center marks a section that is harder than average, with good Class 3 rock-dodging. At medium levels, this part becomes less technical. The pace tapers slightly and, after several turns, another bridge is passed (number four). A chute and a large rock following signal the approach of more technical Class 3 rapids. When houses begin appearing, the river drops in a left turn, and you'll pass bridge number five at Cambridgeport. From here to the next bridge (1.5 miles), the way is smooth Class 1-2.

After a small bridge (number six), the Saxtons loops left, picking up speed for Sixth Bridge Rapids in the next right turn. There are two ledges here: the first is smaller, and the second drops abruptly a maxiumum of two feet on the left. Hydraulics follow both ledges, and they can be sharp in medium water. This is the heaviest water on the trip (Class 3). You can avoid most of it by hugging the inside right of the turn. This isn't easy because the current pushes outward. This rapids can be seen from the road if you know where to look, but a steep bank and a sharp turn usually hold the driver's attention. The remaining 1.5 miles to the take-out have some playful chutes and drops for Class 2 canoeists.

Take out on the left, upstream from a white church in Saxtons River. Directly below the church is a bridge, and under it there's a waterfall dropping sharply over jutting rocks that would surely beat a swept-over paddler to a pulp. There is a lumber yard on the right bank here.

The gage is located on the right bank, 130 feet upstream from a covered bridge, 0.8 miles east of Saxtons River, and 3.9 miles upstream from the mouth. It is hard to read in snow and ice cover. The lower end of the staff that runs from the gage house to the river reads 5.0. The portion of the gage appropriate for lower readings is 20 feet downstream, attached to a rock ledge. Don't fall in trying to read the gage.

For a run that is similar to the Saxtons in difficulty try the Sugar in New Hampshire.

Shepaug River (CT)

ROUTE 341 TO ROUTE 47
Trip A

Distance (miles)	Average Drop (feet/mile)	Maximum Drop (feet/mile)	Difficulty	Scenery
7.0	40	50	2-3	Good

TOO LOW	LOW	MEDIUM	HIGH	TOO HIGH	Gage Location	Shuttle (miles)
	0.5	1.0	2.0		Route 47 Bridge	7.0

The Shepaug, along with the Bantam, is one of the earliest to free itself of ice and snow, so it is usually canoed during the middle or the end of March. The upper portions of the Shepaug are sportier than the lower, and in high water they can be very exciting. The stream starts quite narrow and remains so until the confluence with the Bantam. The Shepaug has many islands. Water flow depends on the dam situated at the start. Since the upper riverbed is small, when there is a water release many trees suddenly seem to grow in several feet of water, something to concern canoeists because eddying out becomes trickier when you have to fight your way through the brush. Even though most of the trip is quick moving, there are no rapids that require scouting, although fallen trees can be a real hazard anywhere. The Shepaug is usually most challenging for open boats. A road follows the upper half, but it is usually not in view.

To reach the put-in, turn north on Route 341 near the Route 202 bridge that crosses the Shepaug at Woodville. After a short distance on Route 341, take the first right, which is somewhat obscure, and continue until you reach a stone bridge. Just upstream from this bridge is a small dam that can be run in the center. For the larger dam upstream, you'll need a parachute. Below the bridge, the central canoeable channel is about one boat length wide and the rapids there are quite typical of the whole run — for 75 to 100 yards, standing waves rush over small rocks and ledges. At a gage of 0.5, waves measure about one foot. The current is a bit pushy and disorienting. An island downstream divides the channel and there's an abrupt drop of .5 to 1 foot over a small rock ledge on the right side. The river then turns right and straddles another island. Later there's a series of haystacks. Past the Route 202 bridge is a small pool. On the left, the exit from the pool leads to a fast, twisty passage.

When you see Romford Road close on the right bank, watch for an island full of brush which can sometimes block the way. Pass on whichever side looks best. Beyond, in a right turn, you'll find nice rocky rapids even if you're not looking for them. After the next bridge (Romford Road), the Shepaug turns right and there is a quarter to a half mile of continuous standing-wave rapids that should be good sport for competent boaters and challenging for others. At the end is a brief stretch of quieter water that approaches the Romford Road on the left bank (about 25 feet up the bank); the river then turns right, leading into a real good rapids that lasts about 25 yards. Usually you can run this rapids in the center or right center. The last two sets of rapids are the hardest on the trip, rated Class 3 at a gage reading of 0.5. One possibility for taking out is at the next bridge, which is just above the confluence with the Bantam. Since the drive in is not particularly easy, most people will choose to stay with the river and take out at the Route 47 bridge, three miles farther along.

When the Bantam adds its water from the left, volume increases noticeably and the water is pushier, especially if the level is high. A short distance downstream, a rock cliff on the right marks a good Class 3 rapids. Just after a left turn with several rocks in the center comes another spicy entanglement with water and rocks, followed by another good rapids in a left turn. Next look for a drain pipe about two feet in diameter on the left bank — choppy water is ahead. This last rapids is in a right turn with flat, slab-like rocks on the left bank and dome-shaped rocks in the water; this is just before the take-out at the Route 47 bridge, where Bee Brook enters from the right side. If the gage reads 2.0 or more, there will be few rocks evident in this last section — or anywhere on the trip for that matter.

There is a hand-painted gage on the left, upstream wing of the Route 47 bridge. Open canoes can manage the Shepaug at most levels, although with a fair amount of bailing after the rapids at high levels.

For other similarly rated trips in the area try the Ten Mile or the Salmon.

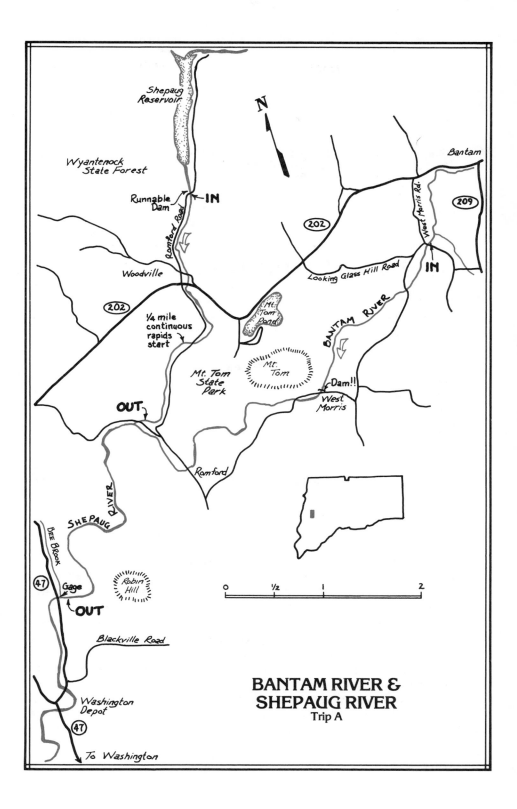

BANTAM RIVER &
SHEPAUG RIVER
Trip A

Shepaug River (CT)

ROUTE 47 TO ROUTE 67
Trip B

Distance (miles)	Average Drop (feet/mile)	Maximum Drop (feet/mile)	Difficulty	Scenery
10.5	18	50	1-2	Good

TOO LOW	LOW	MEDIUM	HIGH	TOO HIGH	Gage Location	Shuttle (miles)
-0.4	0.0				Route 47 Bridge	8.5

The lower Shepaug is a fine early spring river for beginners who want a reasonably uncomplicated Class 1-2 trip. The rapids are mostly standing waves, although a rock pattern can occasionally make matters a bit more complex. The best path is usually obvious. Since this is not a difficult run, you can make the trip at almost any water level. As the level gets higher, however, the current strengthens markedly: beginners take note. Most rapids are less than 100 yards long and they occur between smooth-flowing or pool-like expanses of water. A strong attraction of this trip is its scenery — wooded hills and occasional farmland.

Start this trip where State Route 47 crosses the river, near Bee Brook. The river is fairly wide at this point, and low water levels will reveal scattered bread basket-sized rocks. The Shepaug turns left to start the trip. The river is

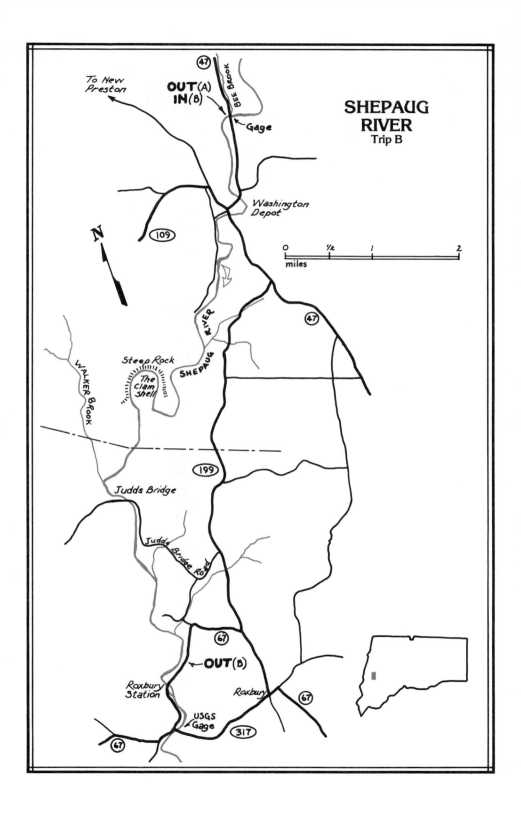

SHEPAUG
RIVER
Trip B

wide for most of its course, with ample room to execute necessary maneuvers. The main channel narrows considerably several times, but this is no great challenge except in high water. At a gage reading of 0, the highest standing waves will be on the order of one foot. Islands frequently split the channel, so canoeists have to choose the best route — usually an easy decision.

The take-out on Route 67 can be made nearly anywhere convenient once the road approaches close to the river.

The gage is painted on the left, upstream support of the Route 47 bridge.

If you would like another trip similar in difficulty to the Shepaug, try the Salmon, which is reasonably close.

Smith River
(NH)

ROUTE 104 TO BRISTOL

Distance (miles)	Average Drop (feet/mile)	Maximum Drop (feet/mile)	Difficulty	Scenery
2.0	90	100	4	Good

TOO LOW	LOW	MEDIUM	HIGH	TOO HIGH	Gage Location	Shuttle (miles)
-0.5	0.5C		1.5C		Cass Mill Road Bridge	2.0

From its origin on Mt. Cardigan, the Smith flows primarily eastward and empties into the Pemigewasset. It is a short, tricky run that can be easily repeated several times in a day. In character, it is typically New England — small and rocky. The section that is usually run is, for the most part, continuous white water, Class 3-4 depending on the level. Because of this, rescues can be difficult. The majority of the rapids offer many paths, although several have just one or two practical routes. Turns and maneuvers must be done with authority, lest the rocks gobble up your boat. Eddies are not uncommon but you must work to get in and stay there, usually crunching over several rocks in the process. A road parallels the river and, although it is no longer a major public thoroughfare, it can still be used for access. It is not plowed. Five miles downstream from this section, the Smith drops through scenic Profile Gorge, which ends in a picturesque waterfall. This portion is unrunnable but you can see it by taking State Route 3A out of Bristol, New Hampshire.

There are several spots from which a trip may be started. Route 104 parallels the river, but a new addition removes it from direct sight. To reach the river, proceed north from Bristol on Route 104 to Cass Mill Road, then turn left. Continue until you see the river. At the crossroads there, a right turn on the old 104 leads to the start, while a left goes to the take-out. You may put in near a bridge crossing the river, or continue until you find a more agreeable location. Old Route 104 parallels the river and intersects new 104, where a trip may also begin. Above this spot, the Smith is mostly Class 1-2. The rapids directly upstream from the old 104 bridge are quite like the rest — a fast current that's pushier than it looks and many rocks to avoid. A small pool at the head of this rapids permits a convenient start, where unused muscles and hangovers may be dealt with in safety. Directly under the bridge, the pace slackens slightly. Below, there's a big rock in right center before the Smith turns hard left. The approach to this turn is filled with haystacks and hydraulics; a heavier rapids follows in the turn, with a two-foot drop into a hole on the right and a smoother tongue in the left center. After 30 feet of straight, fast water, a line of rocks juts out from the right bank, pointing to a chute on the left. Then there's another chute on the right if you miss a rock in the center; it won't move, so you have to. Next on the menu is a fast staircase rapids whose steps measure 1 to 1.5 feet, and are carpeted with turbulence. The Smith continues in this manner, making it necessary for boaters to stay on their toes, thinking and acting fast. A detailed list of difficulties would be too much to remember.

In one right turn just above Cass Mill Road, you'll find a short, tricky spot. As the Smith rounds the corner, the riverbed narrows and gains momentum to pour over a rock blocking the center; then it falls into a menacing-looking hole. In the approach, upstream in left center, is another drop over a small rock. Enter this rapids in the center and let the pillow on the rock throw you left or right, usually left. Probably the second toughest, this rapids is a short Class 4. Immediately below, an island divides the Smith into a larger right channel and a smaller left channel. If you don't look closely, the island is easy to mistake for the left bank. The left way around this island is definitely easier than the right, although three-quarters of the way there is a rock directly in the middle. Going right of this rock is easier, but the current forces the boat to the outside and into another rock downstream, requiring a Z maneuver. In medium or high water, this latter rock will not cause so much trouble.

The right channel around the island can be the toughest of the trip. Approach it cautiously. It has three abrupt drops, several boat lengths apart, and is a comparatively straight shot if you set up properly. The first and third drops are 2.5 to 3 feet; the second is somewhat smaller. Approach in the center for the first, move right for the second, and back to center for the last. The whole thing is 20 or so yards long. Even if you run it correctly, you may have to do some impromptu boat-patching, like putting your stern back. At low levels, it is judged too much of a boat-breaker to try. At the downstream side of the island is Cass Mill Road Bridge; it sits offering a nice view of the whole mess.

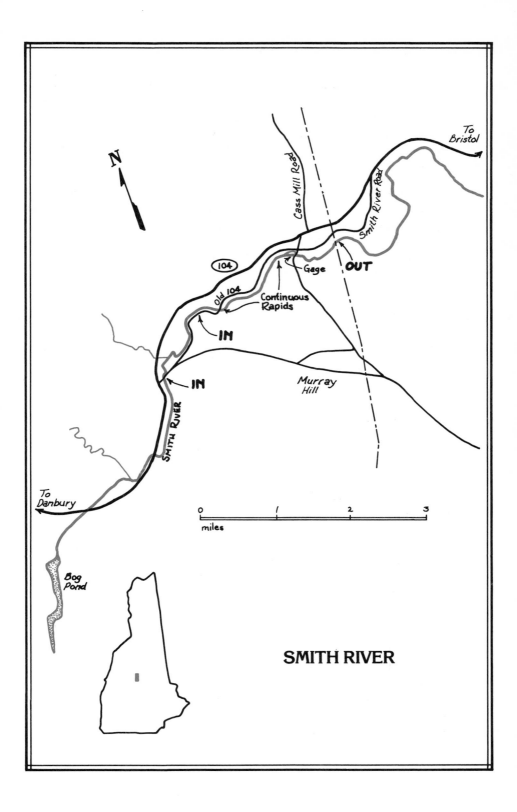

N

To
Bristol

Cass Mill Road

Smith River Road

104

Old 104

Gage

OUT

Continuous
Rapids

IN

IN

Murray
Hill

SMITH RIVER

To
Danbury

Bog
Pond

0   1   2   3
miles

SMITH RIVER

Below the bridge, the Smith turns left and in a bit there's a friendly/ unfriendly double hydraulic in the left center. A snaggle-tooth dead center complicates an otherwise straight approach. At low water the first drop-hole combination is two to three feet and the second is two feet. If you choose left, paddle hard so the not-so-friendly aspects don't pull you back in for a friendly hug and kiss. The rest of the trip is typical Class 3. Take out under a set of overhead power cables.

A hand-painted gage is located on the left, upstream wing of the Cass Mill Road Bridge. There is a U.S.G.S. gage in Merrimack County on the right bank in Hill, 1.5 miles upstream from the mouth and 1.8 miles southwest of Bristol. This gage is very difficult to find and so is not of much use to boaters.

For another nearby river similar to the Smith, try the Mad.

Souhegan River
(NH)

GREENVILLE TO WILTON

Distance (miles)	Average Drop (feet/mile)	Maximum Drop (feet/mile)	Difficulty	Scenery
3.5	50	65	2-3	Good

TOO LOW	LOW	MEDIUM	HIGH	TOO HIGH	Gage Location	Shuttle (miles)
0.5	1.0	2.5C	2.5		Old Powerhouse	4

The Souhegan River meanders in southern New Hampshire, moving north by east. There are two distinct parts to the section described here, a harder upper Class 3 and an easier Class 2 lower stretch. Your main difficulties are over once you've canoed one mile down from the put-in. Enough rocks sprinkle rapids to make things interesting but not especially technical, with two or three exceptions. Most rocks are small, and are covered in medium or high water. At these levels difficulty rarely exceeds tough Class 3. Because of its southern location, the Souhegan is as a rule one of the first New Hampshire rivers canoed in the spring. So, early in the year ice shelves on the banks frequently extend into the river, creating dangerous situations. Route 31 follows the river, although the lower part is more isolated and prettier than the upper.

Start your trip on the Souhegan at the sparse remains of an old powerhouse — which can't be seen from the road — about one mile north of Greenville on Route 31. The former powerhouse is on the right side of the river, and can be reached via an old "paved" road which is partially blocked by a cable where it joins Route 31. The road slants diagonally toward the river. Looking upstream from the old powerhouse, one sees a narrow riverbed with more rocks than below. At the put-in, the Souhegan is 40 to 50 feet wide with a steep-faced left bank. A fairly straight section occurs directly below the old powerhouse, with a rock outcropping on the left bank. Difficulty here depends on the level; however, it should hardly ever go over Class 3. At lower levels, rock patterns are visible; at higher levels there are standing waves. The rapids are generally straightforward, with large rocks on the sidelines. About 300 yards below the put-in, the rocks thin out a bit. In another 100 yards you'll meet the next set of rapids, followed by a shallow section in low water. At the end notice a large boulder in the right center. To the right of this rock the river is about as wide as a boat. The main channel here, naturally enough, is on the left or left center. Then move center, since there are several more rocks on the left side below. This is one of the more interesting rapids of the trip. The outrun is easy. You'll see Route 31 high on the right bank downstream. After more easy rock-picking and haystacks, one passes into a left turn amid larger rocks. A small island appears where the main channel on the right side is narrow and fast. It may be difficult to notice the island as you go whipping downstream.

Shortly comes a narrow left turn. Stay on the inside: this is the entrance to the most difficult rapids (Class 3+ in high water). After this turn, the current is on the left. It races madly 50 to 75 yards toward a group of rocks on the left, through which there is a twisting, boat-width channel. At higher water levels, the right side also become feasible. Quickly after this group of rocks come two ledges, spaced about 10 to 20 feet apart. A big boulder sits in the middle at the start of the first ledge. After a fast runout, then a pool, the Souhegan turns right in the distance. Turn to turn measures 200 to 300 yards. In the vicinity of the ledges, the current follows an S-curve. The ledges can be paddled either on the right or on the left side depending on water level. Both ledges slope and drop about 1 to 1.5 feet, with hydraulics and other nasties following.

From the pool below to the Route 31 bridge crossing, typical Souhegan rapids are in order. A sharp right turn leads to the bridge itself. Just under the bridge on the left, upstream side is a playful hydraulic you might want to try and catch; there's another on the downstream side.

Below the bridge calmer water prevails, and the easier section begins — Class 2, occasionally with more difficult rapids to spice up the afternoon. At a gage reading of 1, this section is rated LOW for open boats paddled tandem and DULL for hard-nosed closed boaters. At this level, typical haystacks measure up to one foot.

Just before a left turn, and five feet upstream from a large boulder in left

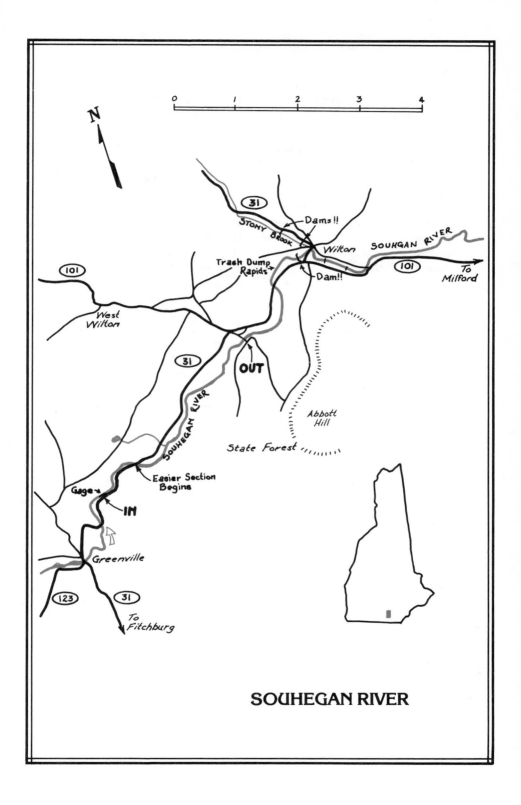

SOUHEGAN RIVER

center, is a 6-inch drop over a rock wall spanning the river — Class 2 rapids. After the left turn there is a nice pool.

Shortly thereafter is another left turn with a trick ending. A rock dead center and rock ledge extending from the right narrow the path. The turn is sharp and fast.

After the Souhegan passes a large gravel pit set back on the left, the river accelerates, turning left very sharply. There are large rocks on the right bank and more below where the current pushes hard against the rocks. This is a tricky rapids if you are inexperienced. In a right turn, a half mile or so downstream, there's a chute that ends with a touch of turbulence. Farther down there's a large boulder on the left; then an island divides the river into narrow channels. This part of the river is away from the road — as deep a wilderness as you'll see on the trip. There are beaver dams, and canoeists should look out for dams that might block the river — this is reported to have happened on several occasions. To the right of one of the islands there is a 1- to 2-foot drop at the start of the channel. Take out by a small bridge that crosses the river.

This trip can continue to the town of Wilton itself. If you decide to travel onward, you get to run Trash Dump Rapids, which is next to . . . guess what! You may also portage a dam just on the outskirts of Wilton.

There is a hand-painted gage on the concrete wall, next to the old powerhouse at the put-in.

Another nearby river — very short, but a little more difficult — is Stony Brook, which joins the Souhegan in Wilton.

Stony Brook
(NH)

ROUTE 31 TO WILTON

Distance (miles)	Average Drop (feet/mile)	Maximum Drop (feet/mile)	Difficulty	Scenery
2	65		3	Fair

TOO LOW	LOW	MEDIUM	HIGH	TOO HIGH	Gage Location	Shuttle (miles)
0.0	.5				Route 31 Bridge	2

If spring has just arrived and you are interested in a short, spirited gambol down a new piece of river, try Stony Brook. For years people have been driving by it on Route 31 on their way to the Contoocook saying, "That little stream looks really interesting." Well, it is. It has some continuous Class 3 water, many 1- and 2-foot ledges, a runnable dam, and a waterfall that spills into a short, narrow, steep-sided gorge — and all within one mile. How's that for concentrated action? Put several feet of water in this stream, and you'll have all you can handle. Stony Brook is also good if you don't want to run the nearby Souhegan twice in a day. Both open and closed boats can find sport here at the appropriate levels.

Since Route 31 parallels the river, you can start a trip almost anywhere. For a short (1-mile) exciting trip, start at a bridge approximately one mile north of Wilton on Route 31. A small stream enters here near the right, downstream side of the bridge. Stony Brook is also fairly wide here and relatively calm.

Several minutes' paddling over uneventful water brings you to a broken dam that can be run only on the left. This chute is narrow, fast, and C-shaped, curving always to the right. The left side of this channel is bounded by unforgiving rocks which approach dangerously fast. A fairly large pool follows where swimmers can be reunited with their boats. Immediately below are numerous ledges under a railroad bridge. The ledges are Class 2 at low water, whereas the dam is Class 3 under the same conditions.

Next are four sets of rapids; each is a series of ledges that drop anywhere from one to three feet. You may want to scout a few of these in high water. In low water all can be run by open boats, but there will be some keel-scraping in the process.

Shortly after the last of these rapids there are a red wooden bridge and some structures on either shore. Immediately below this point a rock ledge blocks the river, and the current funnels into a tight channel on the left which plunges precipitously six to eight feet down into a short but extremely narrow canyon. This drop is at least Class 4, even in low water, and most people will do well to walk around it. If your bow were to get caught on the way down or in the washout, you could spend the weekend in your boat wedged in tight by the current. Portage on the right, but ask permission first if people are around. One hundred yards of Class 2-3 water follows, with Route 31 close by on the left.

The first dam in Wilton is near a large brown building on the left bank. The second is just before a bridge in town. The take-out is anywhere you desire.

For a slightly longer run, put in about a mile farther upstream from the original start. This extra length of river should present no difficulties save for trees in the water, which will definitely be a problem at high levels.

A gage is painted on the right, downstream side of the Route 31 starting point bridge.

For another Class 2-3 trip in the area, try the Souhegan.

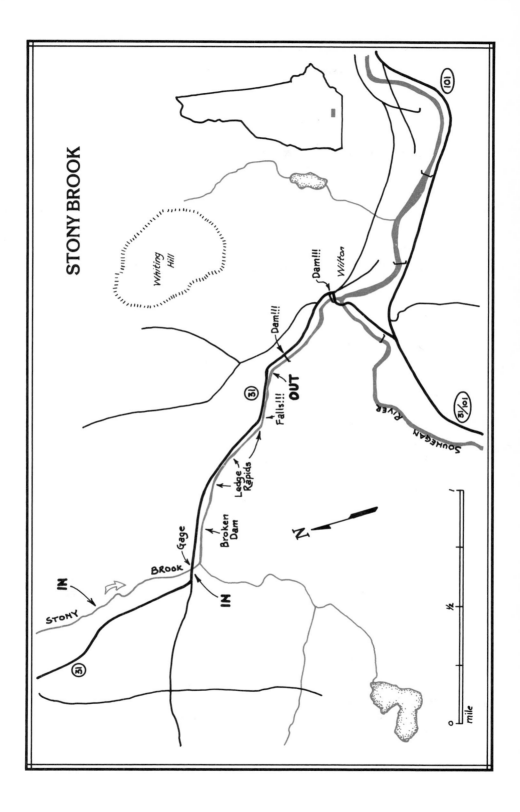

STONY BROOK

Sugar River
(NH)

NORTH NEWPORT TO ROUTE 103

Distance (miles)	Average Drop (feet/mile)	Maximum Drop (feet/mile)	Difficulty	Scenery
2.5	34	40	2-3	Good

TOO LOW	LOW	MEDIUM	HIGH	TOO HIGH	Gage Location	Shuttle (miles)
1.8		4.0			West Claremont	

The Sugar River flows in southwestern New Hampshire, emptying into the Connecticut River at Claremont. The section reported here is Class 2-3, with continuous rapids for much of its length; it is generally removed from heavy civilization. Lots of rocks make an easy technical run, and there's one short stretch of tricky rapids. It is an excellent trip for open boat practice without fear of overpowering rapids, holes, etc. It is also short and easy to repeat in one day. Fairly close are the Black (Vermont) and the Cold (New Hampshire), which are of similar difficulty.

After consulting several road maps and some local natives for directions, put in at the North Newport river crossing just below a railroad bridge and an old dam. If you look closely, you'll see that the middle section of the dam is missing. It is found about one mile downstream. A flood years ago apparently placed it in its present position. At the start, the Sugar is 75 feet

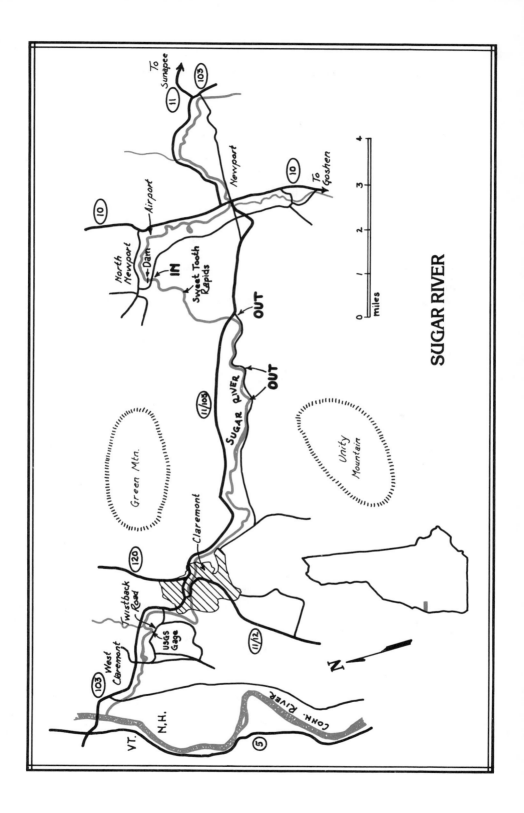

SUGAR RIVER

wide, going right past an old brick building that used to tap part of the river for its power. The river here is Class 1-2, with the small rocks covered in medium water. A railroad bridge is next. Below is an easy Class 2 rapids, then a slightly more difficult one that lasts for some time and passes the middle part of the dam. Again, a railroad bridge crosses the river, and a right turn follows; soon there's a pool and a short Class 3 rapids. Railroad tracks now on the right bank continue to Route 103.

The Sugar widens and grows shallower. A right turn follows opposite a high dirt left bank which is badly eroded. Sweet Tooth Rapids, the toughest (Class 3) on this trip, occurs just before and during this turn. The course, which runs 20 to 25 yards, is complicated by an array of large and small rocks. Two large ones sit in the left center, with lots of smaller ones scattered around. The main channel is right center through narrow chutes, then just to the right of one of the large central boulders. Lower levels demand some fast maneuvering. At higher levels it's more of a straight shot with turbulence. Large boulders also sit on the banks. Alternative routes are possible with adequate water, but decide for yourself. It's a great spot for an inexperienced party to wrap rocks with aluminum or plastic. A pool rests immediately below.

In the next left turn there is a sporty Class 2-3 rapids as the Sugar turns sharply against the right bank, which is cluttered with sharp stones that were used to build the railroad. Calmer water follows, then there's a long Class 2 section past a set of old bridge supports on either shore. The last significant rapids follows shortly. It is a long Class 2-3 rapids that gets harder as it progresses. Near the end the rocks get larger and more numerous, necessitating some maneuvering. This is probably the second most difficult section. An island then divides the river. Both channels are OK but can be shallow. This marks the end of the interesting water; it's only a short paddle to the take-out at Route 103. The trip can be lengthened from here; a turnoff from 103 follows the left bank, creating an easy exit lower down. As the river approaches Claremont, however, it slackens in pace and the scenery deteriorates.

The most difficult part of the Sugar is finding the gage. The concrete house is in Sullivan County, on the right bank, 0.2 miles downstream from Redwater Brook in West Claremont and 2.4 miles upstream from the mouth. It is in a section of the river that loops away from the road. The most useful outside calibrated staff is on the left bank, directly across the river on a rock ledge. To get there, cross the Sugar on a small bridge near a blinking yellow light about 3.0 miles from the mouth. Drive straight until you hit Twistback Road, then turn right. Proceed until the river comes very close by a red house (.25 to .5 miles from the bridge). The gage can be found by stomping through the woods for 25 yards down to the rock ledge. When last seen, the staff had no numbers indicating foot markings, although it did have staples. Take the top foot marking as 4 feet and count downward for readings. Higher levels are measured on a staff attached directly to the gage house itself.

Suncook River
(NH)

PITTSFIELD TO N. CHICHESTER

Distance (miles)	Average Drop (feet/mile)	Maximum Drop (feet/mile)	Difficulty	Scenery
6	20	30	2-3	Good

TOO LOW	LOW	MEDIUM	HIGH	TOO HIGH	Gage Location	Shuttle (miles)
4.3				10	N. Chichester	8

The Suncook is a medium-sized river that runs into the Merrimack just south· of Concord, New Hampshire. It flows east to west and presents an excellent opportunity for beginning canoeists in both open and closed boats. The first couple of miles are mostly Class 1-2, while the lower portions of the trip are more spirited. The rapids, once they start, are uninterrupted in several locations. The whole trip runs through an attractive valley broken only by an occasional house or bridge. No rapids need be scouted, although one relatively hard one lurks in a blind right-hand turn. Most rapids have a scattering of small rocks and a sprinkling of larger ones. You can zig-zag in and out of these natural obstacles or just plot a relatively straight course, as you prefer. Since the Suncook is relatively wide, there is little danger of fallen trees blocking the entire river, except where islands shrink the channel. As the water level increases, the rocks disappear, the current speeds up, and the run becomes one of negotiating a heavy current and standing waves.

Start the trip from Pittsfield, New Hampshire. In the center of town there's a large dam with a turbulent outflow. Where old Route 28 crosses the river, you'll see a leather factory on the left bank. Take the road (Joy Steet) beside the factory until the river comes relatively close for a put-in. You will probably have to traverse a short wooded area before getting to the river. At this point the river is 40 feet wide in low water, with a steady current. Surroundings are mostly meadowlands. The trip is uneventful for a short time, until the Suncook makes one particular right turn. The river then enters a relatively isolated and scenic section; difficulty is Class 1-2. This section ends in a sharp right turn, which marks the beginning of some more interesting water. From this spot to beyond Webster Mills, the current is faster and the rapids more challenging. The next left bend starts a sharp S-turn where the rapids are Class 2-3. The runout is several hundred yards long, with a good current and ample opportunity to practice rock dodging. A set of power lines crosses overhead halfway down. The current continues down to the next right turn, which is sharp, with the heaviest single stretch of rapids on the trip (Class 2 in low water, 3 at higher levels). There are rocks on the outer, left bank and a series of standing waves in the center channel. The inside is the easiest route. The heavier-than-usual water lasts down to the bridge at Webster Mills.

Below this bridge the river widens and loses depth and navigation is difficult if the water is low. At a gage of 4.3, this section is downright annoying. An island also divides the river; either side is OK.

The Suncook soon enters a low-lying section where the river is narrow, the current is smooth, and there are many tight turns — not especially difficult. The trip ends at the next bridge in North Chichester, which is not much of a town anymore. The old bridge here is now blocked. If you'd like a little extra thrill, there's a 2-foot runnable dam, 75 yards below the bridge. It is easiest to run in the middle, but each boater should make his own decision. If you run the dam, the take-out is on the right bank where Route 28 comes close to the river. From Route 28, take Depot Road to arrive at the take-out in North Chichester. This turnoff is near a variety store and a chapel.

The TOO HIGH gage reading of 10.0 was rated because the river was clearly in flood at this level, although most parts looked runnable. The rapids observed were in the Class 3 range, and the river was very much out of its banks near the take-out. Class 2 canoeists should definitely stay off the Suncook at this level; others will have to decide for themselves.

The gage is located on the left bank, attached to a tree, shortly downstream from the bridge at North Chichester. It is not maintained anymore, so it will decay with time.

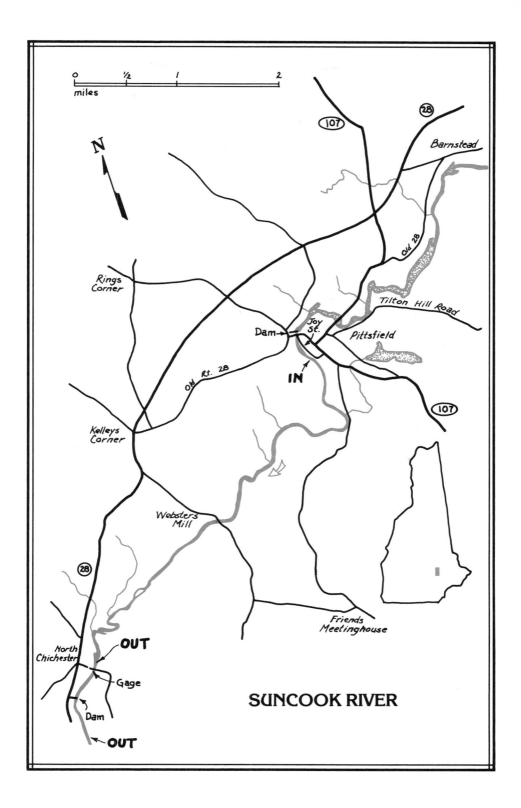

SUNCOOK RIVER

Swift River
(NH)

If veteran boaters were to pick five of the best white water rivers in New England, the Swift would surely be on everybody's list. Arising in the White Mountains and flowing generally eastward along the Kancamagus Highway, it offers one of the most challenging runs in the Northeast. Although only a medium-sized riverbed, even by New England's standards, it boasts some of the biggest, meanest souse holes and choppiest standing waves to be found anywhere. However, its real specialty is rocks and ledges. There are more abrupt drops on the Swift than on any other commonly canoed river in the area — two of them are unrunnable waterfalls. If the water level is MEDIUM or above, the pace is very fast, with little or no interruption between rapids. The pauses between major rapids are generally good Class 3, while most drops are solid 4. In high water, two sets of rapids are rated Class 5. At low or medium levels, the Swift is a long, hard day's paddle, requiring exact boat placement and careful planning. In high water the entire run becomes very difficult, physically demanding, and *dangerous*. If you scout, you'll spend your day walking instead of canoeing — every other drop appears to need eyeballing. Since most of the water comes from snow runoff, the Swift is *cold* as well as mean. So when the Swift is up, most good boaters somehow turn up here. Along with its great charm and challenge, the Swift does have one big drawback — it is an avid boat-breaker. When you boat the Swift, bring along the resin and cloth, the duct tape and winches, because nobody goes home after finishing the river without a personal autograph from mean Mr. Swift. However, when you put it all together, the price of patching is worth it.

The Gorge on the lower Swift. The water in the photo is low — Lynn Williams

Swift River
(NH)

BEAR NOTCH ROAD TO ROCKY GORGE
Trip A

Distance (miles)	Average Drop (feet/mile)	Maximum Drop (feet/mile)	Difficulty	Scenery
4.0	22		1-3	Excellent

TOO LOW	LOW	MEDIUM	HIGH	TOO HIGH	Gage Location	Shuttle (miles)
2.4		3.1			Bear Notch Rd.	4.0

Known as the Upper Upper Swift, this run offers a pleasant trip for those who want stunning scenery and an easy warmup period before the real excitement begins. The first half of this trip is principally Class 1, with many downed trees and meanders. If there is water, the current here is continuous, and the bottom is sandy (even if there is no water). Halfway into the trip, the rapids start, and they too become continuous, reaching Class 3 difficulty in places, at a gage reading of 3.1 and above. The trip ends at Rocky Gorge Scenic Area, where the Swift plunges over an abrupt 10-foot falls. Obviously you should scout the take-out in detail to eliminate *any* possibility of tangling with this drop.

Start your trip where Bear Notch Road crosses the river. If you turn off the Kancamagus Highway, you should reach the river with no trouble. If you

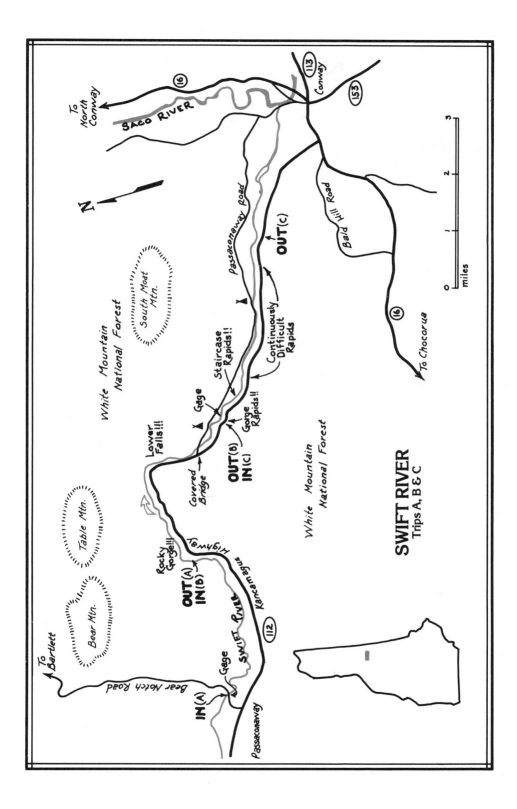

SWIFT RIVER
Trips A, B & C

are trying to come over from Bartlett, however, Bear Notch Road may be blocked by snow early in the season. At the put-in, the river is 60 feet wide and it flows very gently. There are several summer homes. The gage is on the downstream side of the right bridge support. In the distance, the Swift executes an S-turn, and shortly you'll see another summer house high on the left bank. The water is Class 1 and will remain so for quite some time. Be particularly watchful for downed trees, especially in the turns: there are always plenty just waiting to drop in for a bath. The number of angled trees in this section bears silent witness to the power of previous high water.

Although by now it might seem that the whole trip won't get beyond Class 1, rapids do start after a particular right turn; first just standing waves and a noticeably faster current, then one little rocky rapids in a left turn, and there's even another set farther below. The maneuvering is not difficult, and the rapids are spaced out at first. However, these spaces shrink progressively as you move downstream. Rapids up to this point have no outstanding difficulties, so they are not described in detail — you wouldn't remember them anyway.

There is one rapids that does deserve special notice. It is a double river-wide ledge with the two drops separated by 10 to 15 yards. These ledges are the hardest rapids you'll encounter on the trip, and you may want to stop and scout the best route. Several courses are possible, and water level plays an important role. The approach to these ledges is continuous Class 2 at a gage reading of 3.1. You can sight this spot from the Kancamagus Highway if you are looking carefully for it.

Below Double Ledge Rapids, the water is uninterrupted Class 2-3, with the going generally rougher than it is above. The rapids continue right on down to Upper Falls, so it is very important that you recognize the correct take-out spot. The bridge across the Swift at Rocky Gorge is *below* the falls.

The gage for this section of the Swift is located on the right, downstream side of the Bear Notch Road Bridge.

Swift River (NH)

ROCKY GORGE TO GORGE
Trip B

Distance (miles)	Average Drop (feet/mile)	Maximum Drop (feet/mile)	Difficulty	Scenery
3.5	40		3-4	Good

TOO LOW	LOW	MEDIUM	HIGH	TOO HIGH	Gage Location	Shuttle (miles)
0.5	1.3C			3.0C	Gorge	3.5

The Upper Swift is definitely easier than its downstream brother, but it too becomes difficult in high water. When the Gorge gage reads 1.5 to 2.0, most rapids are Class 3 or 3+, and several pass into the Class 4 column. This trip begins in a park-like area and ends in another, so you can be sure of a sizeable audience if the weather is good and Dad wants to bring the family out for a trip. The Kancamagus Highway follows the whole run. Since it is almost always visible, an impromptu exit can be executed without much bushwhacking.

To start, put in below the Rocky Gorge Scenic Area just off the Kancamagus Highway. This spot is also called Upper Falls because the Swift plunges ten feet in vertical falls into a narrow, box canyon-like area that lasts for 100 to 200 yards. This area has been "civilized" by the Forest Service and it swarms with tourists in the summer. The banks are steep, almost vertical, and carrying your canoe down to the river can be quite eventful. If there's snow or ice, the banks make a fantastic sliding board with a big surprise at the end. The rapids start immediately below the put-in, but they are not difficult, in the Class 3 range.

In a left turn, there are some good standing waves, then some large rocks in the middle as the road takes a short break from the river. Another left turn has haystacks and a hydraulic; the next right meander houses a 1-foot drop over a river-wide ledge where several large rocks wait for the next ice age on the left bank. The left side is best here.

The next left has another drop between rocks on the right bank and in the left center — take it straight on. Then more rapids just below, and there's another really good one between large boulders (Class 4). The river then turns right into a heavy Class 4 rapids that has two parts, the second being a 2- to 3-foot drop into a gaping hole with a large stopper punctuating the end. A large rock on the right bank constricts the channel here. A center run gets the worst, or the best, of it depending on how you look at it. The road then comes into view on the right bank. Next, some islands split the channel. They can be run on either side among rocks and whitecaps.

Several turns down, in a dogleg to the right, large rocks appear to choke the flow, a landmark for the start of the final set of major rapids. There are three ledges; each drops one to three feet. The suggested route is to start in the center for the first ledge and move left for the remaining two. Alternative routes are possible, but don't get overzealous, because Lower Falls is waiting for you only 50 yards downstream. Each drop in this rapids is followed by its fair share of hydraulics and haystacks. Without proceeding too far downstream, take out on the right side where you'll find a picnic area full of spectators rubbernecking at the crazy people in the water, just waiting to see someone bounce down the falls. This last rapids may be inspected from the roadside turnoff when you set the shuttle.

Lower Falls drops about ten feet, in stages. It is quite wide compared to Upper Falls. Many people have looked at these Falls, and a number have actually run them, although it takes the right water level and the right frame of mind to do so. Don't look upon running these falls as a challenge you must overcome. Immediately below the falls, the water is fast from the outflow and rather soon a long rapids appears. This rapids is a rowdy Class 3+ at medium levels and vicious 4 in high water. There are several large rocks on the left, halfway down, where you'll find the toughest water. The real problems, however, are the smaller rocks and the turbulence they create. The current is fast and the waves are very choppy, so you must maneuver frequently to avoid being pushed into a broaching situation. This rapids extends around the first right turn, tapering off a bit as it goes. You can scout most of this stretch from the road. To start a trip at Lower Falls is one hell of an introduction to the Swift. At this point it is 2.5 miles to the Gorge, and most rapids are Class 2-3

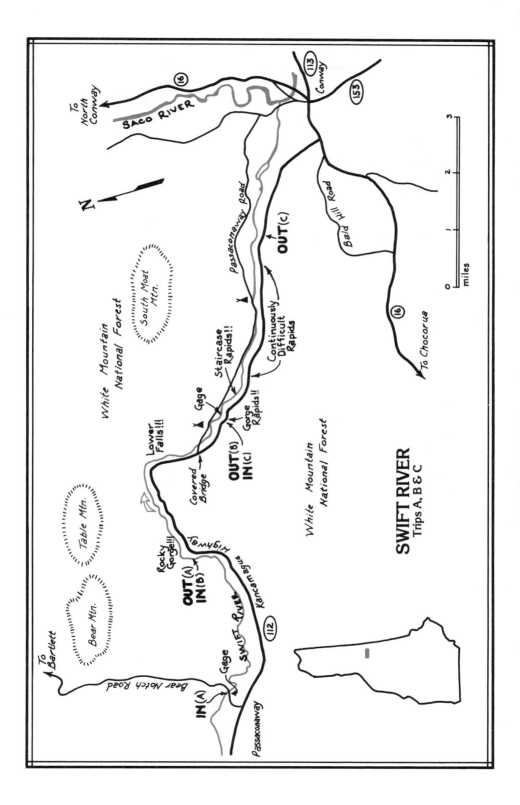

SWIFT RIVER
Trips A, B & C

according to the water level. Along the way, you'll pass a covered bridge and have a little more time to enjoy the scenery. Rapids are mostly discrete, relieved by pools or calmer water. None should require scouting under normal conditions.

Most people will choose to take out before the Gorge, which is a mile below the covered bridge. The Gorge starts just after a rather sharp right turn, although you should not use this turn as a marker because once you are past it you are into the middle of the Gorge, and you may not want to be. Upstream from the Gorge, the road is very close to the river.

The gage appropriate for this section is painted on a rock, on the right bank at the bottom of the Gorge. It is usually visible from the top of the steep bank there. If the gage rock is unreadable because water is flowing over it, you should definitely consider running some other river.

Swift River (NH)

GORGE TO KANCAMAGUS HIGHWAY
Trip C

Distance (miles)	Average Drop (feet/mile)	Maximum Drop (feet/mile)	Difficulty	Scenery
4.0	80	100	4	Good

TOO LOW	LOW	MEDIUM	HIGH	TOO HIGH	Gage Location	Shuttle (miles)
0.5	1.3C			3.0C	Gorge	4.0

If the upper section doesn't get you, the lower one will. Considered one of the most technical runs in New England, this part of the Swift offers challenge almost without pause. After two Class 4 to 5 sets of rapids for openers, the Swift then drops 300 feet or so in the next 3.5 miles, for the most part in discrete jumps. The Kancamagus Highway is close all along the run, so you may terminate the trip anywhere. This lower section is runnable even when the upper section isn't, although under such conditions there will be much boat scraping. A guardrail on the river side of the road occasionally makes parking difficult, but there aren't any parking meters yet.

The start of this trip is most impressive — the Gorge. This is not strictly a gorge, but a particularly difficult rapids that has been given this ominous name. The Gorge is about a hundred yards long, and the total drop is ten to fifteen feet. Canoeists must move around large rocks in a fast, tricky current.

The swiftness of the descent and some car-sized boulders obstruct the view ahead. In addition the Gorge also harbors some rather mean souse holes, as well placed as mischievous sand traps around a green. The approach to the Gorge is a sharp right turn, which is narrow and fast. Many people underestimate this turn, and they end up swimming the whole Gorge before they've even seen it from their boats. Access to this right turn is via a left turn that runs away from the road. Water level means a lot when running the Gorge. Different passages become more attractive at different levels. In low or medium levels, try a zig-zag course on the left side, making good use of the eddies there for the first three-quarters of the way down, then moving to the right side to complete the very last drop. Just upstream from this turning point is a rock with potholes in its downstream side (Turnstile Rock), directly in the middle of the river. Turnstile deflects the current to either side, creating an ugly upstream hole as it does so. Just upstream from Turnstile Rock, one of the many abrupt drops in the Gorge complicates the whole situation even more — definitely a place to avoid flipping over. Below Turnstile Rock at the end of the Gorge a hole extends from the left and tapers into a smooth tongue on the extreme right side, where the only safe passage lies. At its deepest, this hole is about three feet lower than the water immediately about it, and it can be strong enough to hold a boat or a boater. Below the Gorge, the banks are almost vertical and the water is quick but smooth, handy for the recovery work that is inevitably needed. This area is known as the Shirley Siegel Pool. In LOW or MEDIUM water levels, the Gorge is rated a technical Class 4; in HIGH levels, it is a Class 5. Even those who are familiar with it should scout the Gorge every time.

Canoeists who wish to avoid this down payment to the river gods can start in the calmer water below the Gorge. The put-in here is tough because the banks are very steep. The subsequent third of a mile is Class 3-4 depending on water level. There is one sharp drop of two to three feet in this section, and the current is speedy and pushy. And next comes the Staircase.

Lying in wait in a slight left turn, the Staircase is short but very intense, the meanest and most sinister rapids on the trip, as well as an absolute stern-smasher. The Staircase is basically three abrupt drops, about a boat length apart from one another, and it is 15 to 20 yards of concentrated fury. You can run on the left or right. There are numerous holes in the Staircase. One in particular will make a milkshake out of you if you do not run it correctly. If the best route isn't patently clear, stop and scout before running. In high water, the Staircase is a clear-cut Class 5, and a Class 4 even in low; make your own decision on this one.

A shallow rock island just before and alongside of the Staircase separates a narrow rocky channel from the main stream. To avoid the Staircase entirely, take a route to the extreme left of this island. Be alert for a very sharp right turn on the downstream side of the island. If you don't make the turn, you smash directly into a large boulder and end up dropping two to three feet into a hungry hole. Other than that, the going should be easy.

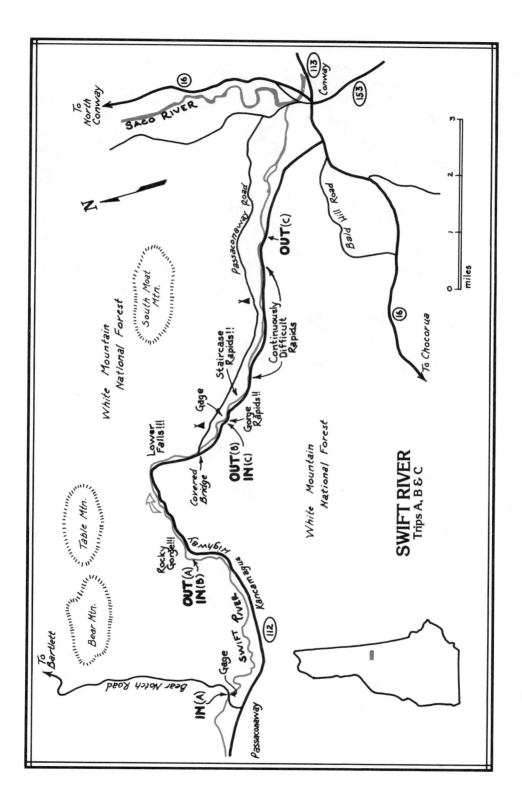

SWIFT RIVER
Trips A, B & C

The approach to Staircase Rapids on the lower Swift — Lynn Williams

Below the Staircase, the Swift goes into its drop, turn, and twist act. It contorts like a bucking bronco for the next couple of miles over one ledge after another. It is impractical to describe all the difficulties in detail — they would be impossible to remember. Rocks are numerous and passages intricate; the pace is nonstop, and, although there are plenty of them, eddies are hard to capture. There are many abrupt drops of two to three feet in heavy chutes and tight curves with powerful water pushing to the outside where there is *always* a rock. The water is extremely turbulent. In MEDIUM or HIGH levels, many hydraulics have the power to hold a boat. At HIGH levels Class 3 rapids can be considered a rest stop. Among the rapids you will run along the way are Screaming Left Turn, Race Course, and Big Rock.

There are two take-outs on the lower Swift. The uppermost is right after Big Rock, by a flat rock ledge that juts out from the right bank to create an abrupt drop. The second is after shallow rapids near a billboard-like sign on the right bank. In both cases, the road is only a short walk. The first take-out is a good place for lunch — the hole created by the rock is fun for playing a hydraulic, unless the water is too big, in which case the hydraulic plays you.

The Swift is mainly a closed boat run, and fiberglass has an advantage over aluminum in that it scrapes and gouges less over rocks. A shorter boat also has an advantage due to the maneuvering required. Lately, a number of people have been running the lower Swift solo in uncovered open boats, and they have been rather successful at it. This is not for everybody; it takes great familiarity with the river and precise boat control, not to mention some luck. In general it is best for inexperienced canoeists to stay off the Swift.

The gage on the Swift is a hand-painted one, on a rock at the end of the Gorge. This rock is on the right side, and you can usually read the gage by standing just inside the guardrail on the downstream side of the Gorge. If the gage rock is underwater, go home, go to lunch, go to another river, but don't run unless you are prepared for a Class 5 trip.

Ten Mile – Housatonic Rivers (NY-CT)

SOUTH DOVER TO GAYLORDSVILLE

Distance (miles)	Average Drop (feet/mile)	Maximum Drop (feet/mile)	Difficulty	Scenery
4.2	12	15	1-2-3	Good

TOO LOW	LOW	MEDIUM	HIGH	TOO HIGH	Gage Location	Shuttle (miles)
			4.6		Webatuck	6.0
			7.1		Gaylordsville	

The Ten Mile is a little-canoed piece of water that could easily be included during a weekend on the Bantam and Shepaug at the start of the spring season. The upper section of this run presents an easy introduction to the more difficult lower portion that joins with the Housatonic. Most rapids are non-technical, with haystacks and a scattering of small rocks. The scenery is pleasant but not spectacular; the river encompasses two states. There is no convenient take-out spot at the end of the Ten Mile trip, so the Housatonic must be paddled. The Housatonic is large compared to most rivers commonly canoed in New England and has two heavy-water rapids.

Put in at South Dover by a bridge and a dam which are just north of Webatuck, New York and State Route 55. Depending on the water level, the dam may be run through a chute on its left side. The water below is not difficult

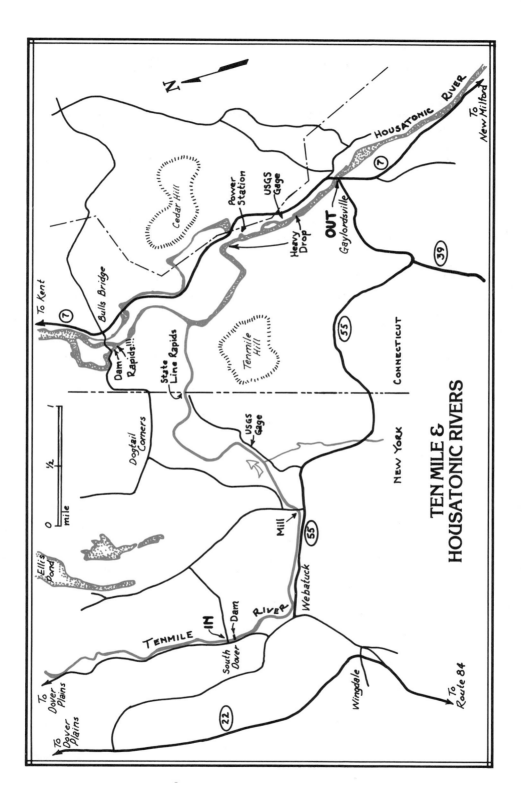

TEN MILE &
HOUSATONIC RIVERS

and the first left turn has a little ripple as Route 55 passes alongside. You'll see a bridge in the distance, and the approach has a few irregularities of laminar flow; otherwise, everything is Class 1-2. After the bridge, the river turns sharply left, passes an old mill on the left bank, and then a broken dam which is easy to run anywhere. After more relatively quiet paddling, the next landmark is a gaging station with its associated overhead cableway. After the gage, the river's pace quickens with a series of Class 2 rapids, and then all is quiet again. A good rapids appears in a right turn where, in lower water, rocks will complicate the path.

Next, after a left turn, a straight section approaches the heaviest drop on the river — State Line Rapids. If the water is high, it is rated a heavy-water Class 3 and it is best inspected from the left bank if you decide to scout. The rapids, which lasts only 15 to 20 yards, is split into two channels by an island and brush pile. The paddler can choose the easier right channel, which may be scratchy or nonexistent in low water, or the more exciting left channel, which funnels and drops into some turbulent whitecaps. Standing waves follow and lead into a pool that has a beach area on the right bank. A farmhouse here is the last structure to be seen on the river. Shortly below the pool, easy Class 2 rapids await. Midway through the pool, the canoe comes under the jurisdiction of the ICC. The rapids that follow are easy Class 3 and extend to the mouth (.5 miles). Standing waves, a few rocks, and a little turbulence are the story here. These rapids are straightforward and easily handled by an open boat at almost any level. There are even short breathing spaces. This is the most interesting section.

Immediately upon entering the Housatonic there is an extended rapids of similar technical difficulty to those just passed, but, since the Housatonic is three to four times larger than the Ten Mile, the water packs more punch. This rapids extends 200 yards downstream and stops in the first left turn where two larger rocks guard the extreme right. Later, when you notice some overhead power cables and a right turn in the distance, head for the right shore to look over the next drop. This drop and the next heavy rapids are rated Class 4 in high water, less in lower. Sprawling across the river, the first rapids presents a variety of faces. In the extreme left center, perpendicular to the current, is a very ugly hole which should be avoided in high water when it could hold a boat or boater. In the right center is a long series of large but regular standing waves that gradually diminish in height over a distance of 25 to 50 yards. This is the preferred ride and is akin to that infamous dipsy-doodle, the roller coaster. At high water, an open boat could easily swamp. You can also sneak by on the extreme right side. Dump out in the middle and you'll set an endurance swimming record getting to shore. A power station is just downstream on the left bank; it releases water from a reservoir high on the left bank.

After passing the gaging station, also on the left, there is a second heavy drop. On the left a ledge sticks out from the bank, while in the center, standing waves flail away at effervescent demons. If the Housatonic is high, these waves are very turbulent and should not be attempted head-on except by the masterful in closed boats. A suggested route is to approach in the left center, just to the right of the ledge, pass the ledge, and then scamper sharply left to avoid the heavier water. Real water-eaters should ignore this last instruction. An open boat could make this maneuver if paddled solo, and possibly tandem. The center portion of heavy water extends farthest downstream, and the left is relatively quiet below the ledge. This rapids can be seen and inspected from Route 7. Then it is just a brief journey to the bridge at Gaylordsville, where a smaller standing-wave rapids (Class 2-3) awaits under the bridge for the huskies who don't mind the carry back.

A trip on these two rivers is not complete without a visit to the Bulls Bridge Rapids. Take Route 7 north to Kent and turn left at a stoplight. After crossing the covered bridge, look both upstream and down. If the water is coming over the dam, it is a most spectacular sight. Water, huge water, is roaring down an impressive chasm at breakneck speed, only to cascade over yet a larger drop. If Dante had been a canoeist, the river Styx would have looked like this. There's only one way you can go. It's also an appropriate spot for training kamikaze kayakers.

The gage on the Ten Mile is on the right bank, 0.1 mile downstream from Deuel Hollow Brook, 1.2 miles upstream from the New York-Connecticut line, and 1.7 miles upstream from the mouth. It can be reached by turning off Route 55 onto Old Forge Road (which is about 0.4 miles from the mill and broken dam). The gage is about a half mile along this road. If the curious care to, they may continue on to the road's end and inspect State Line Rapids. The gage reading of 4.6 is rated HIGH only for State Line Rapids. Those rapids lower down on the Ten Mile are easier at this level.

The gage on the Housatonic is on the left bank, 0.4 miles from a hydroelectric plant, and 0.5 miles upstream from the Route 7 bridge over the river. It is close to the road but may be difficult to see. The HIGH reading for the Housatonic is for the two heavy drops only.

Waits River (VT)

WAITS RIVER TO ROUTE 25B

Distance (miles)	Average Drop (feet/mile)	Maximum Drop (feet/mile)	Difficulty	Scenery
10	38	50	2-3	Good

TOO LOW	LOW	MEDIUM	HIGH	TOO HIGH	Gage Location	Shuttle (miles)
0.4		2.0			Route 25B Bridge	10

The Waits is one of Vermont's northernmost rivers. It is best as an open boat run. Possessing a steady current, it has several large sloping ledges, many sharp turns, an abundance of haystacks, and the smell of cow dung in the air. For most of its course, the Waits can be straightforwardly run, and it is a good place for learning the techniques of solo paddling. In three separate places the upper river is punctuated by ledges that generally require skill to negotiate. The part directly below Waits River, Vermont is decidedly unattractive, although the left bank of the lower part is as scenic as any. For closed boaters, the water must be high to generate excitement.

In Waits River, Vermont, near a white church, turn left off Route 25 and proceed 100 to 200 yards to a small bridge crossing the river. The Waits is about 30 feet wide here, flowing smoothly with very few rocks, looking very much like the countryside stream it is. Moving downstream there are many sharp curves, which become trickier as the water level rises. Houses and farms

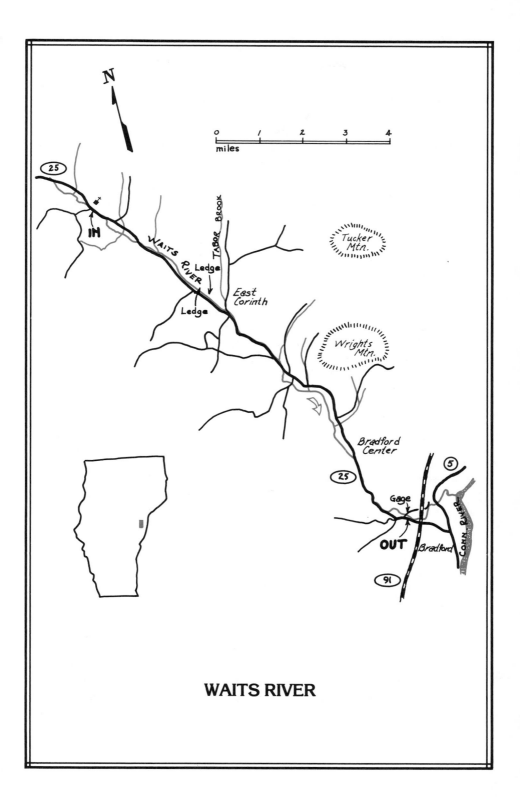

N

0   1   2   3   4
miles

25

IH

WAITS RIVER

TABOR BROOK

Ledge

Ledge

East Corinth

Tucker Mtn.

Wrights Mtn.

Bradford Center

25

5

Gage

CONN. RIVER

OUT

Bradford

91

WAITS RIVER

populate the left bank, as does a collection of old junk cars. Waves and small rocks continue for about 1.5 miles to the first Route 25 bridge crossing, where the approach has a nice chute. Below, the pace increases to form continuous Class 2-3 rapids — fast current, rocks, crosscurrents, and sharp turns. Next, look out for a large, dug-out gravel pit set back behind some trees on the right bank: the next left turn has a small ledge extending from the right. A surfing wave follows. The extreme left side avoids the 6-inch-to-1-foot drop.

A short distance downstream, in another left turn, is a more extensive ledge. Extending across the entire channel, it can be run in the center. There is a rock directly in the center at the bottom of this drop, but it is usually water-covered, so it will push boats aside. The total drop is two to three feet over a distance of ten to fifteen feet. Eddies await downstream on either side and flat water precedes the actual drop. The Waits then broadens out, and you'll see a gas station high on the right bank.

A series of small ledges follows — a 6-inch ledge that extends the width of the river, then the first of two large ledge systems. In a fairly straight section of river, the first system is river-wide and lasts for about fifty yards. It has a short, flat spot midway down. The whole thing resembles a large washboard. One-quarter of the way down this ledge, a small stream enters from the right side. Different water levels will make different routes more attractive — so look this one over before running. A suggested route starts on the left, angles right to just above where the small stream enters, and then goes down from there. After traversing the brief plateau, the last part drops more abruptly, one to two feet. A rooster tail sticks up in the right center, so pass to either side (the right side being better in low water). The whole ledge angles slightly left to right.

In the distance, the river again drops from sight over another large ledge. Smaller than the previous one but similar in character, this one is also river-wide and has been run on both the left and right sides depending on water level. Look it over. Lower down are some smaller, easy ledges and the remains of an old bridge. The road reappears, and then comes the concrete bridge at East Corinth. This is the end of the most difficult section.

From East Corinth to the take-out at Route 25B, 6 miles or so, the Waits is Class 1-2. Directly below the bridge at East Corinth several small streams add their water, deepening the river. If the upper section is TOO LOW, then this lower section can be LOW. Past the first bridge, the passage is quite easy. The valley is attractive, the left side particularly so. The steep-sided left bank occasionally pulls back its rouged mantle of conifers and broadleafs to expose a more picturesque rocky face that sometimes extends to the depth below.

In the approach to a small, dark-colored bridge there is a 1- to 2-foot drop over a rock ledge in the middle, with easier channels to the right and left. Below the bridge there is a long Class 2 rapids with the road close to the left side. After this, the rapids become fewer and consist mainly of haystacks. Below a green bridge are some islands that divide the river into many channels, all of which are passable except for tree hazards. The difficulty here is mainly Class 1. Take out by the Route 25B bridge, just upstream from Interstate 91.

A hand-painted gage is located on the right, downtream side of the Route 25B bridge.

For a trip of similar difficulty that is close by, try the Ompompanoosuc.

Wardsboro Brook (VT)

WARDSBORO TO WEST RIVER

Distance (miles)	Average Drop (feet/mile)	Maximum Drop (feet/mile)	Difficulty	Scenery
4.5			3-4	Good

TOO LOW	LOW	MEDIUM	HIGH	TOO HIGH	Gage Location	Shuttle (miles)
0.0					West River Confluence	4.5

The Wardsboro is a small, seldom-canoed tributary of the West River. Its season is limited — only a couple of weeks unless there are heavy rains. But when the water is up, the Wardsboro presents a demanding run. The current is continuous for the entire run, as are the rapids. It is definitely harder than its nearby cousin, the Winhall — the rocks are larger, the gradient is steeper, and the rapids are generally more complex. As with the Winhall, a road parallels the entire trip, so impromptu exits and entrances are possible.

The basic character of this stream changed dramatically after the flood of 1975. Huge volumes of water passed down all the rivers in the area, altering not only their courses but their complexions as well. In the clean-up period that followed, bulldozers reportedly manhandled the riverbed, so that many parts now look like an uneventful sluiceway. The Wardsboro is far from dead, however, and it still holds a surprise or two. Since there are so many rapids on the Wardsboro, only the highlights are discussed in detail.

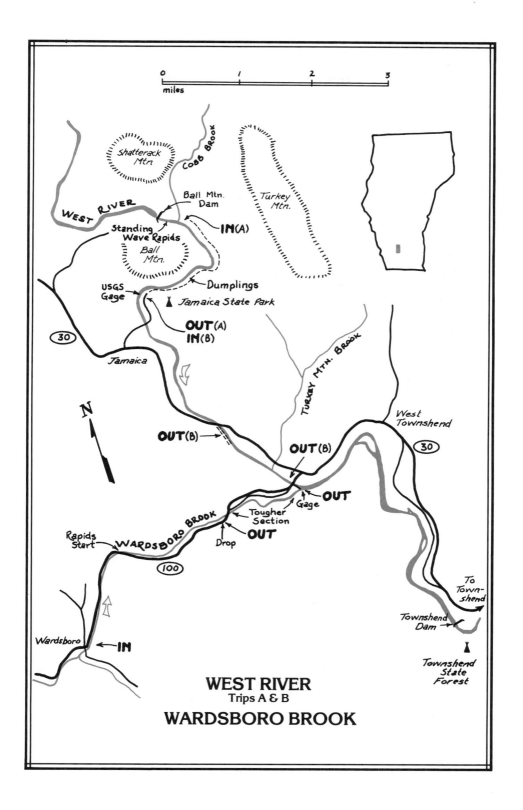

miles
0 1 2 3

Shatterack Mtn

Cobb Brook

Turkey Mtn.

WEST RIVER

Ball Mtn. Dam

Standing Wave Rapids

IN(A)

Ball Mtn.

USGS Gage

Dumplings

Jamaica State Park

OUT(A)
IN(B)

30

Jamaica

N

Turkey Mtn. Brook

West Townshend

OUT(B)

OUT(B)

30

OUT

Gage

Tougher Section

OUT

Rapids Start

WARDSBORO BROOK

Drop

100

To Town-shend

Townshend Dam

Wardsboro

IN

Townshend State Forest

WEST RIVER
Trips A & B
WARDSBORO BROOK

Several starting spots are possible. One is in the town of Wardsboro, just as Route 100 turns away from the river. Another is by a bridge 1.2 miles downstream from Wardsboro Center. From the town to this bridge, the river is an uninteresting Class 2-3, although this section would be useful for a warmup.

Shortly below the bridge, the river turns right and the difficulty begins. Stay on the inside of this turn at first, then move quickly to the center to align yourself for a series of drops among some boulders. Paddle hard since the water is turbulent throughout. A section of numerous rocks follows where maneuvering through tight channels is necessary. One narrow channel should be scouted: there are a number of nasty rocks and holes that may be unavoidable depending on water level. This particular stretch occurs when the channel pinches as it bears left around what used to be an island. This rapids also tends to change from year to year.

Knife Edge Rapids is about 100 yards long and consists of two parts. It is narrow, fast, and requires good boat control. The approach begins in a right turn, and the rapids then continues through the next left turn. The first part is a boulder patch that stretches all across the river. Depending on the level, there are runnable courses both on the extreme right and left. The right channel is probably better in lower water levels, but it ends in an abrupt, narrow drop next to a large rock ledge on the right bank. The lower part of Knife Edge is a sloping three-foot drop, straightforward except for two rocks that run longitudinally with the current and stick up like knife edges in the right center. Passages exist between them, or to their left. Between the upper and lower parts, the current is fast and the eddies are tricky because of complex crosscurrents. Knife Edge can be seen and scouted from the road.

Just upstream from the Route 100 bridge, near the end of the trip, there is a long and possibly heavy rapids. This one lasts for several hundred yards, with an S-shaped course and numerous rocks and hydraulics. It ends with a splash as the river pours over a 2- to 3-foot drop, which is magnified as the river runs higher. The fast runout leads into another 100-yard-long rapids directly below the Route 100 bridge. The run to the bridge goes around an island which you probably won't recognize as such, especially in low water. In high water you won't care. This rapids is a strong 4 in high water, less at lower levels. It should probably be scouted if the water is high. A big eddy forms on the left, upstream side of the bridge. Since this rapids look like a Japanese rock garden in low water, one questions that it could generate any excitement. The 2- to 3-foot drop seems like a mouse step. Put three feet of water over it, however, and the mouse roars.

A trip could end here at the Route 100 bridge, although a mile remains until the confluence with the West. However, that mile is a little harder than what has been paddled so far. Route 100 vanishes from view at this point, and a smaller dirt road, the access to a number of houses, roughly parallels the river. Banks are sometimes extremely steep and scouting may be difficult. Myriad riverbed rocks and several islands complicate things. The pace is fast, the maneuvering is demanding, and downed trees can cause real problems. If the water is up, this section should be reserved for the experienced. You can also run this section when the upper section is too low.

The river in this lower part is very broad in places. There will be a great deal of turbulence if the water is high. However, at a gage reading of 0.0, open boats could negotiate most rapids, since difficulty is a technical Class 3. Those who choose to run tandem will do well to have paddled together before.

There is one spot, though, where everyone should exercise caution. At this point the Wardsboro flows around an island with the main channel on the left side. This channel is very narrow, and there are a series of ledges, drops, and rocks that must be dealt with for a successful passage. A large boulder sits on the left bank near the end, and it takes some effort to miss it. Following this boulder, the boater must immediately paddle around a tight S-turn, trying to avoid the right bank at the end. Even at a gage reading of 0.0, this rapids is rated Class 4. At higher levels the right side of the island will open up as a possible route, but you should judge this for yourself. You can spot an overhead power line below — but only if you are looking at the sky. This rapids cannot be seen from the road.

The rest of the way to the West is relatively easy, but still fast and sassy. At the confluence, two bridges make a convenient take-out.

There is a hand-painted gage on a rock on the right shore, between the two bridges at the take-out. It is *estimated* that a gage reading of 2.0 would present a challenging run for good closed boaters. At a gage reading of 0.0, the upper portion of the Wardsboro is TOO LOW; however, below the lower Route 100 bridge the river is runnable even at this low level.

West River
(VT)

BALL MOUNTAIN TO SALMON HOLE
Trip A

Distance (miles)	Average Drop (feet/mile)	Maximum Drop (feet/mile)	Difficulty	Scenery
2.5	40	50	3	Good

TOO LOW	LOW	MEDIUM	HIGH	TOO HIGH	Gage Location	Shuttle (miles)
		7.2C/6.2	7.2		Jamaica State Park	2.5

One of the principal waterways of southern Vermont, the West has been the site for important national slalom championships. Although it is blocked by two large flood-control dams, there are still several parts of the West that can present some sport, especially for open canoes. In its watershed are two smaller but exciting streams — the Winhall and the Wardsboro. Since the West is dam-controlled, scheduled water releases in late spring and fall permit boating when nearby rivers are dry. The rapids on the West are mostly straightforward, although one stretch may require scouting if you haven't seen it before. Competent boaters will find several good spots to play, and intermediates will still find a challenge.

Starting a trip from Ball Mountain Dam is not the easiest thing to do. There is an old railroad bed that has been converted into a one-lane Class 3 road, which runs from the parking lot of Jamaica State Park up the left side of

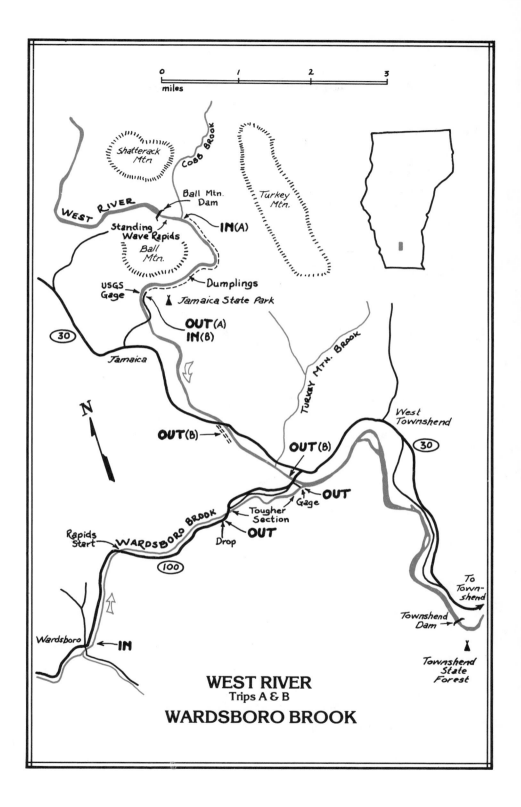

0 1 2 3
miles

Shatterack Mtn

Coab Brook

Ball Mtn. Dam

WEST RIVER

Standing Wave Rapids

IN(A)

Ball Mtn.

Turkey Mtn.

USGS Gage

Dumplings

Jamaica State Park

OUT(A)

IN(B)

N

30

Jamaica

Turkey Mtn. Brook

OUT(B)

West Townshend

OUT(B)

30

OUT

Gage

Tougher Section

WARDSBORO BROOK

OUT

Drop

Rapids Start

100

To Townshend

Wardsboro

IN

Townshend Dam

Townshend State Forest

WEST RIVER
Trips A & B
WARDSBORO BROOK

the river almost to the dam, stopping just short of Cobb Brook. In the past, there has been much disagreement between park personnel and boaters over the use of this road. The access situation seems to change every year, so it is something to be aware of if you are contemplating this route. The portage over Ball Mountain Dam from above is also difficult, but your reward for the trek is the opportunity to run a short section which has the heaviest water on the trip, some large standing waves directly below the single discharge tube of the dam. The West is about 50 feet wide here, with Ball Mountain rising 900 precipitous feet on the right.

In a quarter mile or a little less, near a small, grassy island in the middle, the river broadens, narrows down, and falls over more rocks, to be divided again by a truck-sized boulder in the center. On either side of this obstacle are more boulders — the right side is sportier. Rocks like jellybeans line the sides. Next, Cobb Brook enters from the left, with its final descent in the form of a foamy white staircase.

From Cobb Brook to the Dumplings, the West is for the most part continuous rapids of Class 3 difficulty at a gage of 7.2, and a little less at lower levels. There are countless rocks, waves, hydraulics, and 1- to 2-foot drops, even an occasional hole. This section can be run by open boats — it is not overwhelmingly difficult, although a rescue could be because of the continuous current.

Somewhat below Cobb Brook, where a rock face forms on the right bank and the river turns gently left, the frequency of rocks picks up as does the excitement, at least for a short stretch. Below, the water is flatter, but it still moves as the West then makes a wide looping right turn where there are Class 2-3 rapids, depending on the level. A rock line extends out almost halfway from the left in the turn itself, creating a nice hole on the downstream side. After another left turn, the river then drops gently into a large pool-like area where a huge rock face decorates the right bank. Straining, one can almost see the Dumplings farther downstream.

Appearing like huge blobs of cooking dough left by some giant baker to harden in the summer sun, the Dumplings easily present the most formidable obstacle on the river. However, there is a clear passage through their midst. As you approach, the current speeds up and larger rocks appear on the banks. As you see the Dumplings, look for a huge boulder in the center and pass just to its left, then turn diagonally right and stay close to the inside of this turn as a smaller rock blocks the lower left center channel (mainly in lower levels). As you make this right turn, the passage opens up to view. Pass between this first large boulder and another, lower down on the left. Once past this lower boulder, turn sharply left to enter a fast chute that terminates with a large rock extending from the right bank. The end of this rock supports a fairly hefty hole that can be used for enders and pop-ups. You can avoid the hole by paddling in the left center of the channel. Once set up, the Dumplings run is a regular, but narrow, S-curve among truck-sized boulders and turbulent water.

Looking downstream at the Dumplings. The passage starts to the left of the large rock in the center — Ray Gabler

Looking downstream at the 'S' shaped channel through the Dumplings. The flow here is about 700 CFS — Ray Gabler

At higher levels, other routes open up, but only inspection will determine if they are feasible. The Dumplings is rated a tough Class 3 or an easy Class 4 at a gage reading of 7.2. It looks more fearsome than it actually is. The outrun is fast, with standing waves and large rocks scattered throughout for a good length. There are even some strong eddies that nimble boaters may want to pop into. Once past the outrun from the Dumplings, you won't meet anything difficult before the take-out at the Salmon Hole in Jamaica State Park. The most difficult thing there is waiting in line for the john so you can change your wetsuit in relative comfort.

For a single expert boater in an open canoe, this whole section offers an excellent challenge, especially if the discharge is around 1,500 CFS. At this level, a tandem crew would be taking a long bath, although if the pair were experienced they could have a good time. Be warned, though: at this level rescue of boats and boaters is tough.

With the cooperation of the Army Corps of Engineers, there are two yearly water releases on the West that are planned in advance to fall on weekends. These weekends are usually coordinated by the Boston Chapter of the Appalachian Mountain Club. One weekend is usually in May, and the other is in October. If you decide to go for the October release, be prepared for a crowd that will rival those at a Washington's Birthday Sale; however, things are not all bad — the local Community Church of Jamaica usually puts on a supper fit for a king.

The gage on the West is located in Jamaica State Park on the left shore near the Salmon Hole. However, the external staff read by boaters is directly across on the right side of the river, attached to a rock ledge. So, get into your boat, paddle across the pool there and read the gage to determine if you should paddle back. The gage is underwater at levels of 10.5 and over. It is also sometimes hard to spot because of debris.

West River (VT)

SALMON HOLE TO ROUTE 100
Trip B

Distance (miles)	Average Drop (feet/mile)	Maximum Drop (feet/mile)	Difficulty	Scenery
3.5	30	40	2-2+	Fair

TOO LOW	LOW	MEDIUM	HIGH	TOO HIGH	Gage Location	Shuttle (miles)
	6.2	7.4			Jamaica State Park	3.5

This lower portion of the West is easier than the upper stretch; it is a pleasant paddle at almost any water level. With a larger cross-section, the lower part is shallow in places and in low water it is a bottom-scraper. This broadness also allows enjoyable canoeing for open boats even when the upper part is too high. Although there are well-defined rapids, none are difficult or require scouting, and a straightforward approach and run will suffice. Also, it is possible to avoid all the heavy water by canoeing the sidelines. The road is fairly close to the river almost all the way and there are many signs of civilization, yet the valley is not totally unattractive. For those who care to, the trip may be extended to Townsend Dam, where the reservoir offers boating facilities for those less interested in white water.

Begin the trip just inside Jamaica State Park at the Salmon Hole, which is a pool-like expanse that's flanked on its left side by a sand beach. There are also a paved parking lot, many drive-in campsites, and an outhouse, complete with flush plumbing. (Ah, for the real wilderness.) Rock-lined on the right side with the main current nearby, the Salmon Hole reportedly was the scene of an early Indian battle — nowadays, the only battles are for parking spaces. At its downstream end, on the left, is a low island and below, there's a Class 1-2 rapids that leads to an iron bridge. Under the bridge is a good Class 2-2+ rapids, rocky in low water and stuffed with waves in high water. After a pause, there is a short, intense rapids with heavier waves (2 feet at a gage of 7.4), another pause and then there's a short tricky stretch where the current is forced away from the left bank into a center channel by a rock ledge. In high water this tickler becomes rather angry. The far right is calmer, and an eddy lurks below on the left.

Next, a sweeping right turn houses an extended series of small hay-stacks and rocks where the right bank is actually a large island. The right-side channel of this island is narrow but passable. Be on the watch for downed trees. After a period of quieter water there is another long easy rapids terminating in a left curve. At the end of this turn there's a hole with hydraulics in the center and right. These can be avoided by staying left. The lower part of the island also ends just upstream.

The beginning of a trailer camp on the right bank marks the last rapids before the Route 30 bridge. This one is 50 to 75 yards long, rocky in low water, with two-foot waves at a gage reading of 7.4. It can be run almost anywhere. Since most of the sporty rapids are exhausted, the boater may decide to take out at the bridge and run again or, alternatively, continue down about 1.5 miles to the Route 100 bridge. For those wishing to continue even farther, a rat-like maze called the Corkscrews is the only interest. Here the river is divided into many channels by islands, and the sharp turns between are good exercise in Class 2 boat handling. Take out on a road that follows close to the left side above the dam, or continue to the beach area of the reservoir.

For camping in this area, there is the Jamaica State Park for which a camping fee is charged (more for out-of-staters). There is also the Winhall Brook Camping Area near the Winhall River and an area near the Townsend Dam.

The gage on the West is located in Jamaica State Park on the left shore near the Salmon Hole. However, the external staff read by boaters is directly across on the right side, attached to the rock ledge. So get into your boat, paddle across the pool and read the gage to determine if you should paddle back. The gage is underwater at levels of 10.5 and over. It is also sometimes hard to spot because of debris. The flow in the two sections described here is entirely dam-controlled.

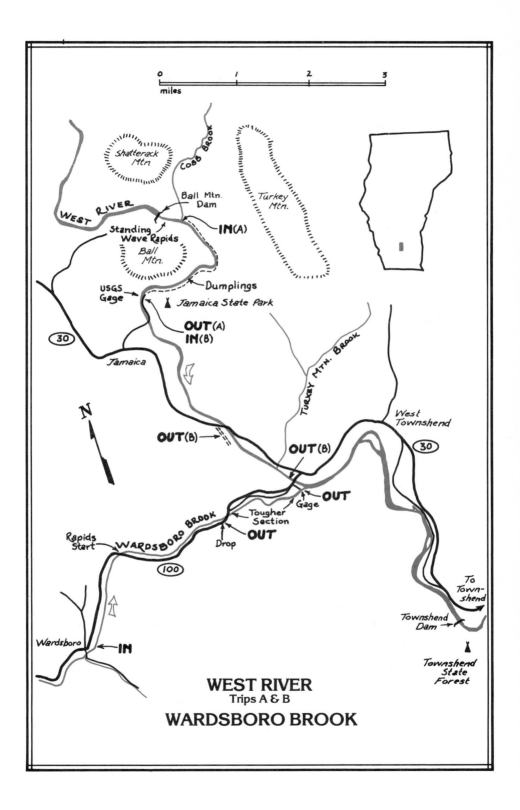

WEST RIVER
Trips A & B
WARDSBORO BROOK

North Branch of the Westfield (MA)

WEST CUMMINGTON TO CUMMINGTON
Trip A

Distance (miles)	Average Drop (feet/mile)	Maximum Drop (feet/mile)	Difficulty	Scenery
6.2	33	50	1-2	Fair

TOO LOW	LOW	MEDIUM	HIGH	TOO HIGH	Gage Location	Shuttle (miles)
0.0		2.0 4.0			Route 9 Bridge Huntington	6.0

The North Branch of the Westfield gently winds its way down from the Berkshires, first crossing Route 116 and then following along Route 9. The trip described here is good for beginners: there are hardly any major difficulties, and the road is always near. The difficulty will generally not exceed Class 2, and then only in high water. The riverbed is narrow at first, but it gradually widens as the trip progresses.

A convenient meeting place and put-in spot for this section is across from the Berkshire Snow Basin Ski Area, where the river is rather small. From here to below the iron bridge on Route 9 near Cummington, the river is essentially Class 1-2 at medium levels. The only difficulty lies shortly below the starting point in the form of a small ledge which is best run on the left in low

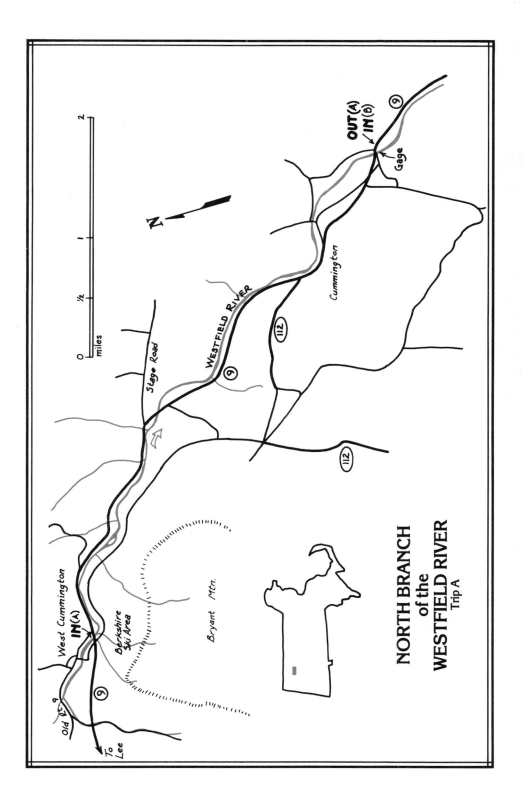

**NORTH BRANCH
of the
WESTFIELD RIVER**
Trip A

water. The channel here is slightly S-shaped, and there's a more abrupt drop over the ledge on the right side (6 inches). In medium water, the ledge should be easier, although the hydraulic should be a little stronger. As the trip progresses, water volume increases, but difficulty does not. Route 9 follows the river for the entire trip and crosses the river twice. Be aware that the trip's start as shown on the map is on old Route 9.

The take-out is by the iron bridge on Route 9 near Cummington. There is a hand-painted gage on the right, downstream side of this bridge.

The Huntington gage reading is on the West Branch of the Westfield, and it is given because this gage is a permanent U.S.G.S. one. The Huntington gage will only be a rough indication of what the North Branch is doing because they are in different watersheds, but a rough correlation is better than none.

North Branch of the Westfield (MA)

CUMMINGTON TO CHESTERFIELD GORGE
Trip B

Distance (miles)	Average Drop (feet/mile)	Maximum Drop (feet/mile)	Difficulty	Scenery
7.2	36	60	2-3	Good

TOO LOW	LOW	MEDIUM	HIGH	TOO HIGH	Gage Location	Shuttle (miles)
0.0	1.0	2.0			Route 9 Bridge	5.4
2.0		4.0			Huntington	

This trip on the North Branch is probably one of the most pleasant to be experienced. Moderate rapids, delightful scenery, and an isolated stretch of river all greet the paddler on a crisp spring morning. And, if your friends show up on time, then you have the start of a good day.

This trip is ideally suited for open boats since the water is seldom overwhelmingly high. The rapids are mostly straightforward, yet it takes a fair amount of finesse to negotiate them with grace. Once Route 9 leaves the river, the boater enters a section that is isolated from all roads, and this part contains the main rapids. If the water is low, the run will be highly annoying, as you'll be scraping over one rock after another.

One good starting point for this trip is at an old iron bridge on Route 9 that is south and east of Cummington. There are also other spots on Route 9 downstream from this bridge in the form of roadside turnoffs. At the bridge, the Westfield is about 75 feet wide and smooth flowing. From the iron bridge to the entrance of the Swift River (2.2 miles), the canoeing will depend on the water level. If the level is low, irritating rock-dodging is in order, whereas medium or high water covers these obstacles with easily run haystacks. There should be nothing dangerous or complicated here.

Entering from the left side, the Swift River adds its volume to the Westfield which makes a 120° turn to the right into a section locally known as the Pork Barrel, which drops an average of 40 feet per mile. In low water, a small standing wave appears slightly downstream from the mouth of the Swift, but it disappears at medium levels. At this point, Route 9 leaves the scene regardless of water level. The Westfield then meanders through a relatively inaccessible valley until a road comes close on the right side. This road signals the end of the Pork Barrel region, which has rapids containing 2- to 3-foot haystacks in medium water. Most of the heavier waves can be avoided by careful maneuvering. There are no "interesting" rapids or ones that need to be scouted in the Pork Barrel unless a tree is trying to get into the act. There are several sharp turns, and the current can be strong. This section is rated Class 3 in medium water, and Class 2 in low water. It should present little challenge to the experienced; however, be aware that you are relatively isolated from civilization during this portion of the trip, which can be a problem if you need help.

Past the Route 143 bridge, the river is rather shallow. Between here and the Chesterfield Gorge, there are several standing-wave rapids, the last in a right turn ending in a long pool with a beach area on the right. Take out here since there are no more convenient places before the Chesterfield Gorge. The Gorge itself is sometimes runnable depending on water level, type of boat, strength of party, etc. The most difficult spot requires negotiating a sharp 3- to 4-foot drop followed almost immediately by a complicated maneuver through some rocks. Although the Gorge is not very long, the walls are precipitous, and it is well worth a visit. The entrance to the Gorge is in a very sharp left turn.

To shuttle for this trip, proceed north from the iron bridge on Route 9 for several hundred yards, turn left onto Fairgrounds Road, take the left fork at the fairgrounds, and continue past the Route 143 bridge to an old graveyard, shaded by a large tree, on the left side of the road. There is a private road leading to the river here. Ask permission before using this road for a take-out, though. If, for some reason, you can't use this road, take out at the Route 143 bridge. In any case, be sure that you know where the Gorge is, so you don't venture in unknowingly.

There is a hand-painted gage on the right, downstream side of the Route 9 bridge at the start. The Huntington gage reading is on the West Branch

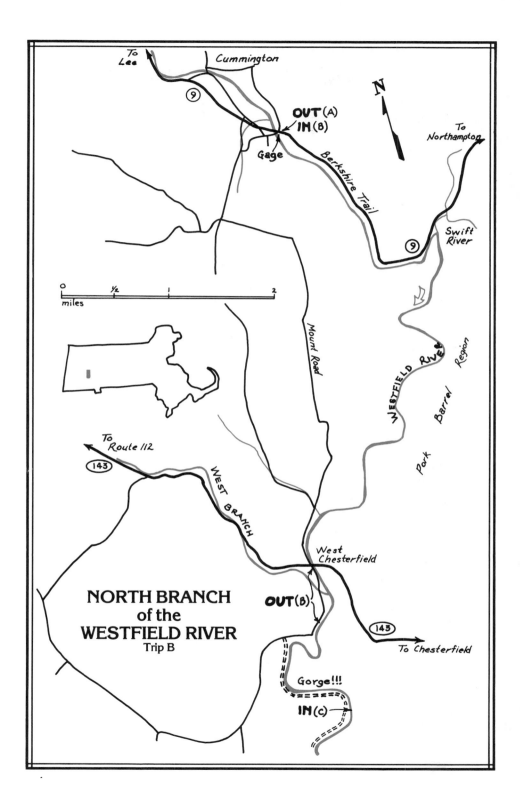

NORTH BRANCH
of the
WESTFIELD RIVER
Trip B

of the Westfield; it is given for a cross-reference and because the Corps of Engineers can obtain its level by phone.

For another river in the area that is similar in difficulty to this trip, try the Middle Branch or the lower part of the West Branch of the Westfield.

North Branch of the Westfield (MA)

CHESTERFIELD GORGE TO KNIGHTVILLE DAM
Trip C

Distance (miles)	Average Drop (feet/mile)	Maximum Drop (feet/mile)	Difficulty	Scenery
9.2	12	40	1-2-3	Good

TOO LOW	LOW	MEDIUM	HIGH	TOO HIGH	Gage Location	Shuttle (miles)
0.0	1.0	2.0 4.0			Route 9 Bridge Huntington	16

This section of the North Branch shows both white and flat water. The rapids are mostly straightforward, and none should require any scouting. However, as the trip progresses and nears the Knightville Dam, the current slows to a stop so the boater must provide some locomotion. The trip is away from main roads, although there is a dirt road that follows the river almost the whole way.

If you are looking for a Class 2 trip, put in below the Chesterfield Gorge — start above it and you'll go up a notch or two in difficulty. A small dirt road off the main road leads to the Gorge and beyond (see map for details). This dirt road is Class 4, extremely tough going especially in wet weather when the road offers more of a challenge than the river.

The rapids below the Gorge are similar in character to those above. There are several noteworthy stretches: the first two merely involve haystacks in turns, whereas the next requires maneuvering through a short section of large boulders. These will be in the Class 2-3 range depending on water level. The last rapids is easy to recognize because the rocks seem to extend across the river. Passageways exist on both right and left. The right channel requires a slight left turn, then a slight right turn. The left channel requires a somewhat sharper right turn. Downstream, an overhead cable signals the end of the interesting water, although one more set of rapids remains. As you approach Knightville Dam, the current slackens, the scenery becomes flatter, and the canoeing is lake-like.

One of the more difficult things about this trip is the shuttle. Proceed back to Chesterfield and take Route 143 north to Worthington Corners, and then take Route 112 south to just above the dam. A shoddy-looking road leads to the river. Alternatively, cars can be left at the dam itself. It is also possible to proceed south on the main road from West Chesterfield, pass the Gorge and Little Galilee Pond, and finally connect with Route 112.

The gage for this section is on the right, downstream side of the Route 9 iron bridge near Cummington. The Huntington gage is on the West Branch, and is given as a reference because the Corps of Engineers can obtain readings by telephone.

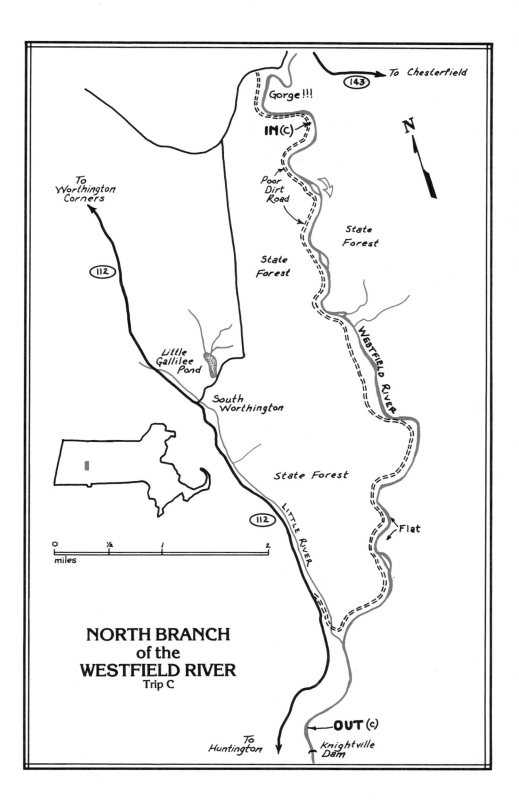

To Chesterfield

143

Gorge !!!

IN (c)

N

Poor
Dirt
Road

State
Forest

To
Worthington
Corners

112

State
Forest

WESTFIELD RIVER

Little
Gallilee
Pond

South
Worthington

State Forest

LITTLE RIVER

Flat

0   ½   1   2

miles

112

**NORTH BRANCH
of the
WESTFIELD RIVER**
Trip C

OUT (c)

To
Huntington

Knightville
Dam

North Branch of the Westfield (MA)

KNIGHTVILLE DAM TO HUNTINGTON
Trip D

Distance (miles)	Average Drop (feet/mile)	Maximum Drop (feet/mile)	Difficulty	Scenery
5.2	17	45	3	Fair

TOO LOW	LOW	MEDIUM	HIGH	TOO HIGH	Gage Location	Shuttle (miles)
	4.5	5.0			Knightville Dam	5.5

This trip on the Westfield covers much of the course where an annual downriver race takes place during the first weekend in April. The race lasts for about ten miles; it demands endurance, strength, the ability to maneuver a boat around rocks, and a little something to keep you warm after a cold swim. Water on this run is supplied by the outflow of Knightville Dam, which is always a factor when this trip is under consideration. Usually the outflow is an advantage, because you can find out ahead of time if there will be a water release and how much. This part of the North Branch is by far the biggest of the four trips described. It is also significantly larger than any other river section of the Westfield system discussed in this book. The rapids offer varied difficulties ranging from heavy water to intricate rock picking. One should also bear in

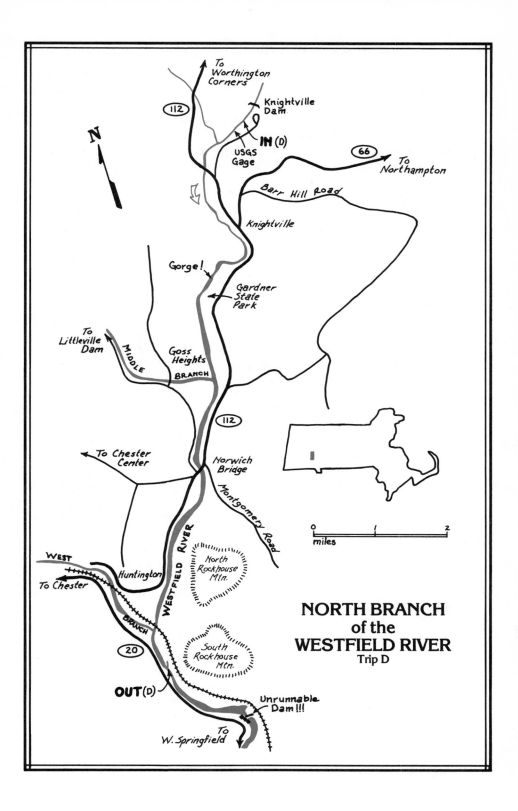

N

To Worthington Corners

112

Knightville Dam

IN (D)

USGS Gage

66 → To Northampton

Barr Hill Road

Knightville

Gorge!

Gardner State Park

To Littleville Dam

MIDDLE BRANCH

Goss Heights

112

To Chester Center

Norwich Bridge

Montgomery Road

WESTFIELD RIVER

North Rockhouse Mtn.

WEST

To Chester

Huntington

BRANCH

20

South Rockhouse Mtn.

OUT (D)

Unrunnable Dam!!!

To W. Springfield

0   1   2
miles

NORTH BRANCH
of the
WESTFIELD RIVER
Trip D

mind that rescues will be difficult because of the large size of the riverbed and a possible strong current. Discharges in the spring of several thousand CFS are not unusual.

Start this trip near a picnic area downstream from Knightville Dam, where the river has plenty of rocks and a fast current. Shortly, you'll come to a small dam (one to two feet) extending the width of the river, which can generally be run anywhere. A hydraulic follows, then an easy ripple just before a right turn. After the first left turn, there is a set of haystacks measuring one to two feet in height at a gage reading of 4.8. Past a bridge and a flat stretch, another wavy rapids in a left turn awaits the eager paddler. Some houses on the left mark a drop through a long rapids, rather straightforward in medium water but rocky in low water, requiring a zig-zag course. The river is very wide here so there's plenty of room for maneuvering. A large boulder on the right marks the end, while a similar one in the middle of the river marks the beginning of still another rapids.

A left turn uncovers a large, picturesque pool with a camping area on the left and a precipitous rock cliff on the right. Downstream, a short gorge section forms. There are two well defined sets of rapids here, both of which have easy and hard routes. The first rapids is about 25 yards long, and is most straightforwardly run on the right where there is open boat fast water. The extreme left has a narrow chute between the bank and a rock. In the middle, the riverbed drops a little more abruptly, creating standing waves which are 2 to 3 feet at a gage reading of 4.8. The center outflow is turbulent, and there are strong eddies on either side.

After a slight hesitation, the river moves on to the second drop. A shallow island divides the river here into a channel at either extreme. The left side is easier and safer, with a small series of tricky haystacks starting a narrow passage. The current is fast, and the eddies are shifty. The right side of the riverbed is a different story. Subdivided still more by several large boulders, the right channel has most of the water pouring through another narrow channel. Pass just to the right of the large subdividing boulder, then move left to avoid two huge, powerful haystacks that are in the middle of the road, spaced about 25 feet apart. Look this stretch over if you haven't seen it before. It is the heaviest on the trip.

The going then becomes easy. The Middle Branch enters and the next bridge announces the beginning of an easy haystack rapids. After this, and passing what appears to be an old bridge or dam support on the right bank, the Westfield starts falling through a boulder patch more rapidly than before. After a brief halt, another boulder patch follows. Both sets of rapids are good Class 3 rock-dodging practice, so they present ample opportunity to pin your canoe around a rock. Both sections can be seen from the road, and during the downriver race in April this area is clogged with spectators. During the course of the day, they usually get some good viewing.

Next, you pass two bridges, the West Branch joins in from the right, and the Westfield turns left. The take-out normally used is found shortly downstream by a picnic area at a roadside turnoff on the right shore.

The gage is located on the left bank, 0.2 miles downstream from Knightville Dam. The external staffs are scattered about, so keep looking. Information about water releases can be obtained from the Corps of Engineers through the Boston Chapter of the Appalachian Mountain Club.

If you would like another river of similar difficulty (although it is a little smaller), try trip *B* on the West Branch of the Westfield. (Trip *A* is a little harder than both the *B* and this trip.)

Middle Branch of the Westfield (MA)

RIVER ROAD TO LITTLEVILLE DAM

Distance (miles)	Average Drop (feet/mile)	Maximum Drop (feet/mile)	Difficulty	Scenery
7.0	43	50	2-3	Good

TOO LOW	LOW	MEDIUM	HIGH	TOO HIGH	Gage Location	Shuttle (miles)
0.0	2.2 3.7	3.0			North Chester Huntington	7.0

When the water is up, the Middle Branch offers a fast, sassy romp that'll challenge open boaters and even give closed boats a good time. Continuous rapids with rocks and haystacks, a good gradient, and backwoods scenery are typical. No rapids need be scouted, although the pace is almost nonstop. A road parallels the river, even though it isn't always seen. The last half mile before Littleville Dam is a little harder than the rest; here the Middle Branch rushes madly through a narrow section, twisting and churning over the many rock patterns.

The easiest way to reach the Middle Branch is to meet in Huntington, Massachusetts, cross the Route 112 bridge over the West Branch, make a 170° left turn on Basket Road, then follow the signs to Dayville Recreation Area. It may also be reached from the north by turning south off Route 143 in West Worthington on River Road, which parallels the Middle Branch. This turn

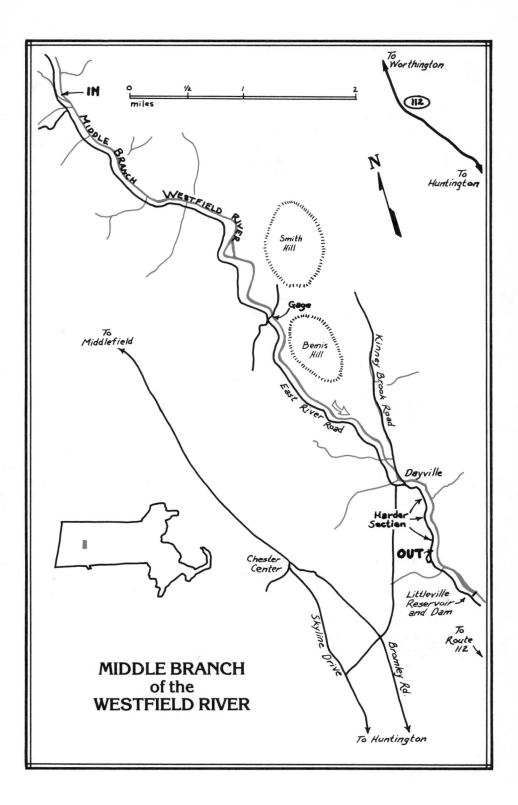

**MIDDLE BRANCH
of the
WESTFIELD RIVER**

IN

To Worthington

112

N

To Huntington

Smith Hill

Gage

To Middlefield

Bemis Hill

Kinney Brook Road

East River Road

Dayville

Harder Section

OUT

Chester Center

Littleville Reservoir and Dam

To Route 112

Skyline Drive

Bromley Rd.

To Huntington

0   ½   1   2
miles

is not obvious. From River Road, turn on Kinne Brook Road to reach the take-out in a parking lot just upsteam of Littleville Dam. This road is usually snow-covered early in the spring; it parallels the lower, harder section.

There is no special spot to start a trip, since the road is always close. However, a suggested spot is some 3.2 miles north of North Chester, where there is a field on the river's right, and a rock ledge narrows the width to less than a boat length. There are several houses also. This is just north of a small bridge which crosses Glendale Brook.

It is impractical to detail all rapids on this trip because there are so many and all are quite similar in character. The difficulty is Class 2-2+ in low water and Class 3 in medium or high. There are countless little ledges, haystacks, and hydraulics, none being particularly dangerous. Almost every turn has a rapids with standing waves and a strong current — none require a great deal of maneuvering. At higher levels, a doubly paddled canoe will take in water.

In a straight section, about one mile above North Chester, there is a 1- to 2-foot dam that can be safely run anywhere. You can recognize it by the line it forms across the river and the smooth water just upstream. A small pool is located below on the left, closely followed by several fast, narrow, winding channels around some islands.

Just before the thriving metropolitan village of North Chester is seen, the Middle Branch turns sharply right, with a 1- to 2-foot ledge preceding the actual turn. This ledge is usually run on the extreme right or left, since there is a hydraulic in the middle. A short, steep-sided gorge full of fast water and a few rocks at lower levels follows. In 2 to 2.5 miles, you'll reach the Dayville Bridge; the level of excitement is reduced in intensity along the way. For those who don't want to continue down the harder section (Class 3), Dayville Bridge is a good take-out.

Below the Dayville Bridge, there is a small pool in low water or a smooth current in medium water. The outlet from the pool is on the right, then the Middle Branch loops sharply left. As you enter this turn, stay on the inside because several large rocks prevent passage on the outside. This section is very narrow and faster than the trip above. This first turn is the sharpest. In low water there are many rocks to avoid, which are replaced at medium levels with a strong current, lots of haystacks, and small hydraulics. Low levels definitely offer more interesting maneuvers. The end of this section and of the trip itself is marked by a low boulder patch in the center with passages on either side; it is covered at a gage reading of 2.5 to 3.0.

There is a hand-painted gage on the right side of the North Chester Bridge support (Smith Road). It can be viewed by looking through the grating of the bridge floor or by walking across to the east side and looking down.

West Branch of the Westfield (MA)

BANCROFT TO CHESTER
Trip A

Distance (miles)	Average Drop (feet/mile)	Maximum Drop (feet/mile)	Difficulty	Scenery
6.0	50	100	3-4	Excellent

TOO LOW	LOW	MEDIUM	HIGH	TOO HIGH	Gage Location	Shuttle (miles)
2.2	3.0	4.0C			Huntington	6.3

To reach Bancroft, go north on Route 20 about 2.2 miles past Chester to Bancroft Road which angles off to the right up a steep hill. If the rocks in the river are covered at Bancroft, one can expect a roller coaster ride below. However, if they are somewhat bare, a fair amount of maneuvering will be called for in several spots. Most rapids are followed by pools. In several places the channel is quite narrow, so fallen trees could be a real hazard, especially at higher levels. A railroad follows the river, but there are no roads close by until the Middleville Road approaches near the end, so the trip is somewhat isolated. The first three miles of the trip are definitely the most exciting; thereafter the rapids are more straightforward, although there is still a strong current. This trip is rated Class 3-4 depending on water level. It is an excellent challenge for doubly paddled open canoes at lower levels.

About a third of a mile below the start a small stream enters from the left, and there follows a fair-sized flat section at the tip of a small island. This is the beginning of a fairly technical section. Both routes around the island are tight; the right side is preferable. About three-quarters of the way around the right side are two narrow chutes, each dropping one to two feet. They are about a boat length apart and can be a boat width wide in low water. They are a bit of a surprise, as they come at the end of a left turn and cannot be seen very far in advance. Afterwards, there is a very small pool and a short, intense rock garden leading to a sharp right turn with a rock wall on the outside. After a brief section of calmer water, the current pushes toward the right bend and the path narrows for a 100-yard rock garden. The river valley is deep and isolated here, far from everything except the railroad. Shortly, the river turns left under a high railroad bridge and continues to fall around rocks and down narrow channels.

In one straight section, the Westfield appears to be closed in by both banks and then drops out of sight. At this spot, two ledges, about twenty yards apart with a total vertical drop of four feet, are runnable in the middle if there's enough water. The last ledge is more abrupt than the first. A large pool follows the last ledge and makes an excellent swimming hole. At a gage reading of 4.0, the hydraulic below this last drop becomes rather mean and ugly, so you may want to scout for the best course when the water level is MEDIUM or HIGH. Another railroad crossing marks the downstream exit from the pool. From this point onward, the rapids diminish in intensity except for a few drops and a ledge (low water) shortly after the approach of the Middlefield Road on the left. This road comes close for the first time after two more railroad bridges are passed. In MEDIUM water, this section is fast moving with many haystacks (Class 3). The approach to Chester is straightforward with no great difficulties. Take out at the first bridge in Chester by the Chester Inn.

The gage is on the left bank, 0.4 miles downstream from Roaring Brook on Skyline Drive. From Huntington Center, go north over the Route 112 Bridge and make a 170° turn onto Basket Road. Follow this road past Broken Dam Rapids until the gage is sighted. It is on the outside of a left turn in the river.

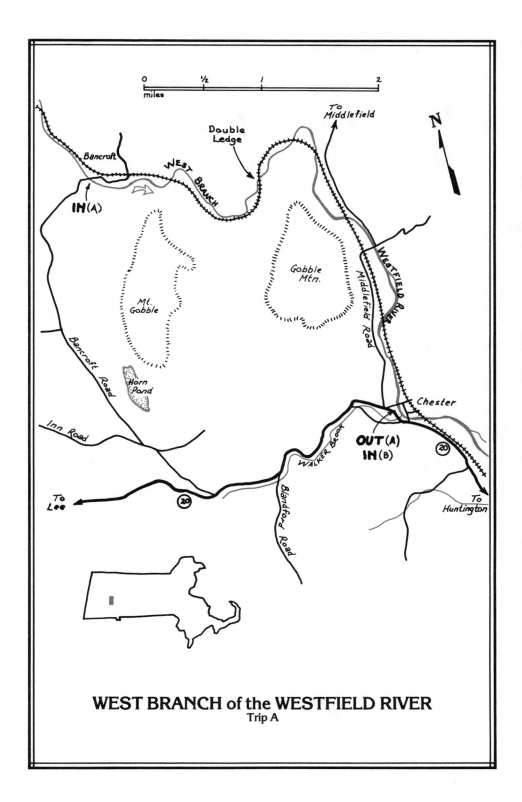

WEST BRANCH of the WESTFIELD RIVER
Trip A

West Branch of the Westfield (MA)

CHESTER TO HUNTINGTON
Trip B

Distance (miles)	Average Drop (feet/mile)	Maximum Drop (feet/mile)	Difficulty	Scenery
7.5	30	40	2-3	Fair

TOO LOW	LOW	MEDIUM	HIGH	TOO HIGH	Gage Location	Shuttle (miles)
2.0	3.0	4.0			Huntington	7.5

This lower section of the West Branch is somewhat wider and has less of a gradient than its upstream counterpart. With well-defined rapids, mostly of the standing wave variety, and many islands which divide the main channel into fast, narrow chutes, this trip presents a sporty run for open boats. It is also a good place to try out old techniques in a new closed boat. The rapids are usually followed by patches of quieter water and occasionally by a stretch where little discernible current can be found. If the upper part is TOO LOW, this run can be LOW. At medium levels, it is rated Class 3. Route 20 parallels this section of the West Branch, and a secondary road also follows alongside on the north side the last half of the trip.

Put in at the upstream bridge in Chester near the Chester Inn. Here the river is over 100 feet wide and rather shallow. The main channel is on the right,

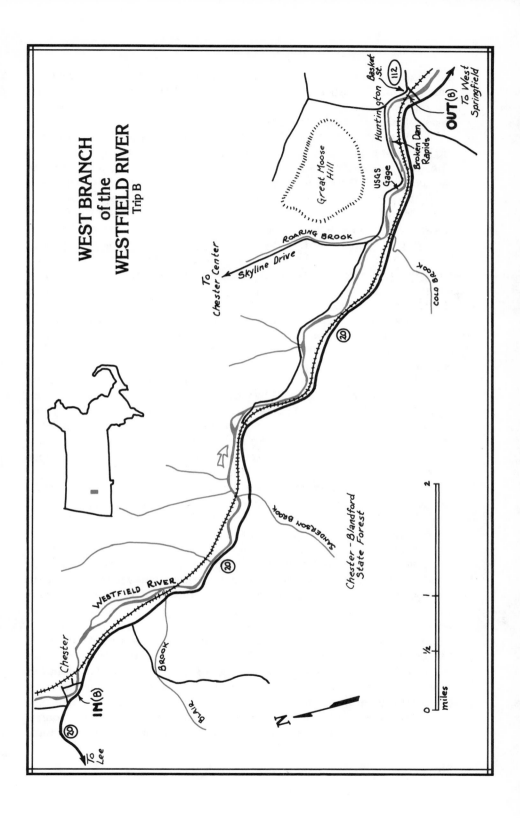

although one can scrape along a short way on the left until one reaches deeper water below. One hundred yards beyond this bridge a 6-inch to 1-foot ledge extends from the left bank to center stream. The right side is clear. Walker Brook also enters in this stretch.

After the first railroad bridge there are several hundred yards of standing waves which are easy to run anywhere; after a brief pause, there are more. These are typical of the rapids on this trip, with just a few being heavier (one or two stretches with rocks to avoid). An island then appears, and either side is OK, although the right channel is livelier. Islands are a recurring theme.

Downstream somewhat, after a little less interesting water, a slight right turn serves as an entrance to another long standing-wave rapids. This one is easily run, as there are few obstacles. You'll meet larger waves at the end. Look for a collection of abandoned cars enhancing the scenery on the right bank, then get ready to enter a nice chute in a slight left turn which has 2-foot waves at the bottom in low-medium water. The river then narrows and passes swiftly in a tight S-turn. Another railroad bridge signals still more standing waves that approach the highway, then turn left. Beyond is a metal bridge, not open to general traffic since both sides are closed. Another island shortly below has a chute on the right side and tricky crosscurrents at the bottom. The left side is straight and fast.

Below the next railroad bridge, which should be run on the left since the right is full of trees and brush, the river turns right and approaches a rapids that is short but requires a bit of maneuvering around several large rocks. Three boulders, two in the right center and one on the left, force the paddler to the extreme right or to left center. A calm spot below is handy for regrouping. A few standing waves also get into the act here. Moving on, the paddler will pass several more islands and more similar rapids.

The next major rapids is Gaging Station Rapids (Class 3). After a slight left turn, notice a concrete wall five feet high and fifteen feet long on the left shore just before the next right turn. The river drops abruptly here, two feet on the extreme left. There is a chute with strong crosscurrents, a large standing wave in the center, and a small drop on the right side. In low water, the approach is shallow, and in medium levels there is a good hydraulic following the drop on the left and center. This is the heaviest rapids yet encountered on the trip, and you may want to look at it if you are unsure of yourself. After a short stretch of quick but calmer water comes another Class 3 rapids (in medium water).

Upon passing a small overhead bridge, look downstream for a large boulder on the left side of the river. This is a signal for the beginning of Broken Dam Rapids, which is rated Class 4 at medium levels. If you want to scout, it is best to do so from the left side by the boulder. On the right side, the main channel picks up speed, dodges a few rocks, and then turns sharply right, only to pile headlong into a stone wall. The current then turns abruptly left, pounding over several large, partially submerged rocks to finally fall into a hole and a series of standing waves. Overall, Broken Dam Rapids has water that is powerful and fast, and the whole rapids is tight. If you run, alignment is critical. Miscalculate just a little, and you'll know what a rag feels like in a washing machine gone mad. At medium levels, only a huge boulder in the left center is visible as a landmark. This rapids is only about thirty yards long and it is more appealing to the closed boater than to the open boater. The take-out in Huntington is shortly downstream, by a white church just before the Route 112 bridge.

The gage is in Hampshire County, on the left bank 0.4 miles downstream from Roaring Brook on Skyline Drive. From Huntington center take the Route 112 bridge over the river and immediately make a sharp left turn onto Basket Road. Follow this road past Broken Dam Rapids until the gage is sighted.

For another trip of similar difficulty in the immediate area, try the Middle Branch of the Westfield.

White River (VT)

ROCHESTER TO GAYSVILLE

Distance (miles)	Average Drop (feet/mile)	Maximum Drop (feet/mile)	Difficulty	Scenery
16.0	11		1-2	Excellent

TOO LOW	LOW	MEDIUM	HIGH	TOO HIGH	Gage Location	Shuttle (miles)
	6.0				West Hartford	16

Draining central Vermont, the White offers many miles of pleasant canoeing through some of the most beautiful countryside New England has to offer. Like the Green, the White is known more for its attractive watershed than for its rapids. Whereas the Green typifies the backwoods, the White characterizes the farmlands of Vermont. Nestled between high, tree-shrouded hills, these tiny farms are tucked among the giant folds of the Green Mountains. Typically a house attached directly to the barn and other buildings for protection against fierce winters, these fertile farms provide the backdrop for the White. The rapids (what rapids there are) consist of Class 1 and 2 water; they need a very heavy spring runoff to make them more difficult. The water itself is as clear as distilled, allowing the lazy paddler to observe fish ten feet below the surface. If it weren't for the cows, of which there are many, the White could provide a natural source of water for a town.

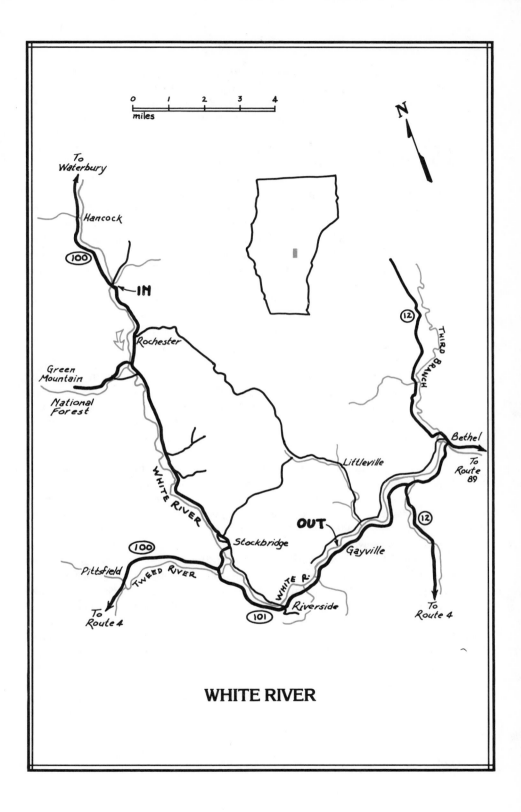

miles
0 1 2 3 4

N

To Waterbury

Hancock

100

IN

Rochester

Green Mountain

National Forest

WHITE RIVER

12

THIRD BRANCH

Bethel

To Route 89

Littleville

OUT

12

Stockbridge

Gayville

100

Pittsfield

TWEED RIVER

WHITE R.

Riverside

101

To Route 4

To Route 4

WHITE RIVER

There are numerous starting and stopping points but a suggested run, easily covered in a day, is from Rochester to Gaysville. Put in by a small roadside turnoff, about 1.5 miles north of Rochester on Route 100. Here the White is twenty to thirty feet wide and flowing with a weak current under a small bridge into a swimming hole. The valley bottom is flat, with hills rising in the distance. In front of the next bridge there is a small ripple as the White proceeds around Rochester. On the right, the bank is steep and attractive, with broadleafs and tall stately ferns. The water throughout this area is Class 1. Below the Route 73 bridge is a pasture and an island, the left side of which has a small chute. Shortly, a small stream, the West Branch, enters from the right adding its waters to the main stream. More Class 1 water follows. The next iron bridge has a little drop, before and after — each is enough to give a bathtub sailboat trouble, but that's all. The river then meanders slowly among wooded hills, past another bridge and a golf course on the left. On the downstream side of the links is a large, emerald-green pool and lower, there's an easy rapids with a few rocks. After a while, old bridge supports appear on either side, and the Route 100 bridge is passed. Below is a long, extended Class 2 rapids. Route 107 is now close on the right. At low levels this section shows many rocks, so the canoeist must move his paddle from time to time.

The next bridge is in a sharp left turn, after which there is a pool and a sportier section. Signaling this section is a long, large boulder in the middle of the river, while below there is a rapids full of haystacks. Next come some more easy ones, then some rocky Class 2s, with old bridge supports standing in the middle while a small stream enters from the right, halfway down. The White then turns left to find easy water again, sprinkled with several more stretches of rapids.

In a left turn, with a rock face on the right, is the heaviest rapids of the trip. Consisting of a series of standing waves (two feet at low water), this one drops into a pool, the outflow of which is in a right turn with rock outcroppings on both sides. An easier drop follows. The heavier upper part can be run almost anywhere, although care should be taken to avoid hitting the fishermen standing in the water. Get too close and you're apt to initiate some fisherman language. The White then reverts to its previous form — calm water with little rapids every now and then. Just upstream from a green bridge, the current swings to the outside of a left turn against some undercut rocks on the bank. The ensuing crosscurrents are a bit tricky. Think left for an easy route.

You can take out at any of the campgrounds that you pass. There are at least two in the Gaysville area and another some two miles downstream. If the boater chooses to continue on to the last one, the dividends are several sets of rocky Class 2 rapids. The road (Route 107) is close by, and a few spectators may gather to watch.

The gage is on the left bank, 700 feet upstream from the bridge in West Hartford. It can be telephoned by the Corps of Engineers throughout the year.

The gage is in a very wide part of the river downstream from the section described here. Because of this, small changes in the gage reading will reflect relatively large changes in the flow.

The Ompompanoosuc River is fairly close by. It offers a run that is somewhat similar in difficulty to the White, although the Ompompanoosuc is the smaller of the two.

Wild Ammonoosuc (NH)

STILLWATER TO ROUTE 302

Distance (miles)	Average Drop (feet/mile)	Maximum Drop (feet/mile)	Difficulty	Scenery
2.0	90	100	2-3	Good

TOO LOW	LOW	MEDIUM	HIGH	TOO HIGH	Gage Location	Shuttle (miles)
3.6	5.0				Bath	2.0

If the Ammonoosuc is too high for comfortable boating, or if you want new scenery, the Wild Ammonoosuc presents a short Class 2-3 run. Starting at Stillwater, where there's a waterfall, the paddler can ease his way down among rocks and standing waves, never venturing far from a road. Although it has a steep gradient, the drop is averaged over the entire length, presenting a continuous fast run where no rapids are difficult or require scouting. In high water the whole trip will turn into one long waterfall.

Put in below the covered bridge and waterfall in Stillwater. Here the river is 50 to 75 feet wide and fast-flowing over small rocks. It continues in this manner most of the way until it joins with its larger brother, the Ammonoosuc.

At one point, notice a gorge-like area; above, the main channel is on the right and the current drops over a small ledge. The water is smooth for a way, then more rocks lie across the path. A center run is OK. The next couple of

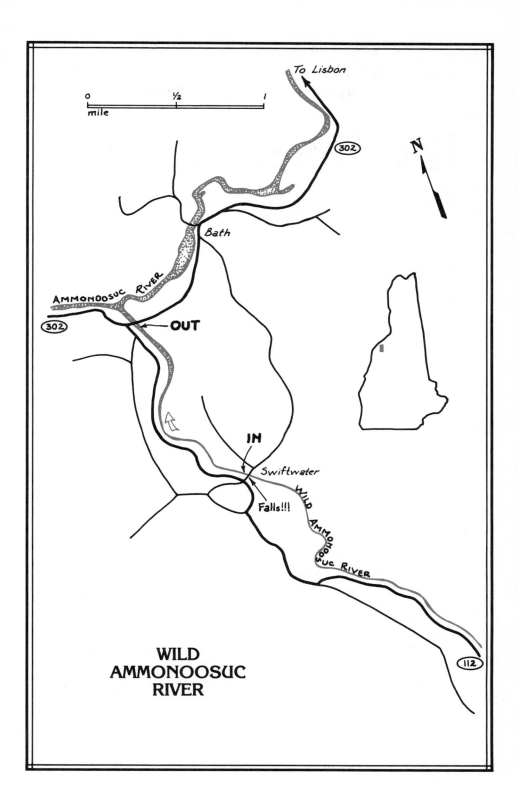

To Lisbon

0   ½   1
mile

302

N

Bath

AMMONOOSUC RIVER

302

OUT

IN

Swiftwater

Falls!!!

WILD AMMONOOSUC RIVER

112

WILD AMMONOOSUC RIVER

hundred yards are cluttered with larger rocks and .2 miles farther, after a right turn with large rocks on the right bank, there's a good chute with rocks. Again, a center run is fine.

The trip may be ended where the Route 302 bridge crosses the river or, alternatively, continued down to the Ammonoosuc.

For a gage, use the one on the Ammonoosuc at Bath. The correlation is not well defined, but it should serve as a good indicator.

Williams River (VT)

NORTH CHESTER TO BROCKWAY MILLS

Distance (miles)	Average Drop (feet/mile)	Maximum Drop (feet/mile)	Difficulty	Scenery
7.5	16	34	2	Fair

TOO LOW	LOW	MEDIUM	HIGH	TOO HIGH	Gage Location	Shuttle (miles)
2.0	3.0				Brockway Mills	7.5

The Williams is a small stream that presents few difficulties, and so it is good for beginners. There is, however, one Class 4 rapids in the middle of the run, and an unrunnable waterfall terminates the trip. For most of its length, the Williams meanders through farmland in Class 1-2 fashion, with the most difficult obstacle being a downed tree or a cow. The river is seldom far from a road, and there are numerous bridge crossings along its length.

Start this trip just below a grist mill and a small dam in North Chester. There is a grassy area and a bridge crosses the river here. A trip could also be started in Gassetts, about 4.5 river miles upstream, but high water is needed for this stretch, which is very narrow. At the grist mill, the Williams is about two boat lengths wide and Class 1 in difficulty. Below, the river turns right and flows through the downtown North Chester-Chester area, which is mainly on the right bank. Farmland is on the left. Along the way, pasture land and a lumber yard are passed.

It should be noted that cows frequently are free to roam in the river and one should take care not to excite them. For some reason, boaters are of great interest to bovines and they often follow a moving canoe by running along the bank. As long as they stay on the sidelines, you have no problem.

The Middle Branch of the Williams enters from the right shortly below Chester. At the confluence there is a nice, deep pool and the river subsequently widens. For the next two miles, typical rapids are Class 1-2 as the Williams moves lazily along. After this distance has been covered, there is a left turn where there is a rocky (low water only) Class 2 rapids with a small bridge at the end. Several hundred yards below this bridge are the stone pillars of an older bridge with another Class 2 rapids in between. Mark this spot mentally, because it is only one mile to Ledge Rapids (Class 4).

As you are making a right turn where there is a house (hard to see) back from the left bank, look ahead for a rock ledge pushing its way out from the left bank. This is the signal for Ledge Rapids. Immediately upon spotting this, head for shore — there's a rather difficult drop ahead. As the ledge moves out, the current is forced to the right, speeding up accordingly to drop abruptly three feet over a jagged edge. The current then turns slightly left, dropping two to three feet more in the next fifteen to twenty feet, and finally runs directly into a boulder in the river. Along the way, there are several angled hydraulics that are quite strong. Most paddlers in open boats should portage this one — easiest on the left. About 100 to 150 yards below, the Williams turns left into a rocky Class 2-2+ rapids, and then, shortly, right again, where more rocks complicate the path, especially at the end. When the right turn has been made, a covered bridge is seen and the most difficult rapids are over. Ledge Rapids cannot be seen conveniently from any road. It could be run by a competent closed boater. A railroad parallels the right side here. From the covered bridge to the take-out, the paddling is again calm and relaxed.

Notice, as you set the shuttle in Brockway Mills, that directly under the bridge there, the Williams drops over an impressive waterfall. *The take-out should be a suitable distance upstream to avoid any chance of being swept over.*

The gage is in Windham County, on the left bank, 25 feet upstream from the bridge in Brockway Mills and 4.6 miles upstream from the mouth. The externally calibrated staff for levels up to 3.5 feet, however, is on a slanted rock ledge directly across on the right bank.

For another trip of similar difficulty try the Cold. For a slightly harder trip, try the Saxtons.

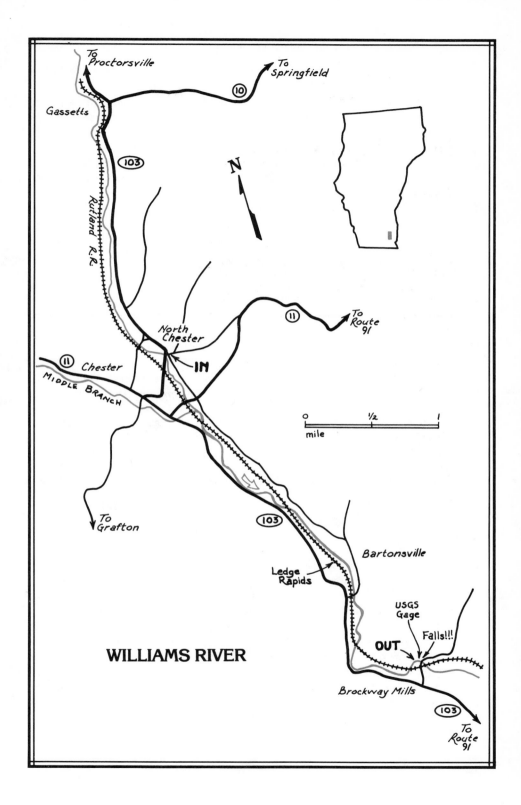

WILLIAMS RIVER

Winhall River (VT)

GRAHAMVILLE SCHOOL TO LONDONDERRY ROAD

Distance (miles)	Average Drop (feet/mile)	Maximum Drop (feet/mile)	Difficulty	Scenery
4.5	62	100	3-3+	Fair

TOO LOW	LOW	MEDIUM	HIGH	TOO HIGH	Gage Location	Shuttle (miles)
0.5	1.0	1.5 2.0C			Route 100 Bridge	4.5

The Winhall is a rather small stream that can offer an exciting ride in high water and a challenging Class 2-3 run in medium or low water. Because it has a small watershed, the Winhall possesses a relatively short season — normally it's up for only a few weeks in early April. The current is continuous and the rapids nearly so. There is a healthy assortment of rocks, but in medium or high water the turbulence is most noticeable, and maneuvering is to avoid hydraulics and haystacks. In low water the rocks necessitate a moderate amount of maneuvering, and a solo paddler running a big aluminum canoe will find himself very busy. There is only one set of rapids that should be scouted, and it is easy to avoid. A road passes alongside almost all the way, so most of the river can be seen before running. The Winhall enters the West River above the Ball Mountain Dam. It is not practical to list all rapids, since there are so many and they are so similar.

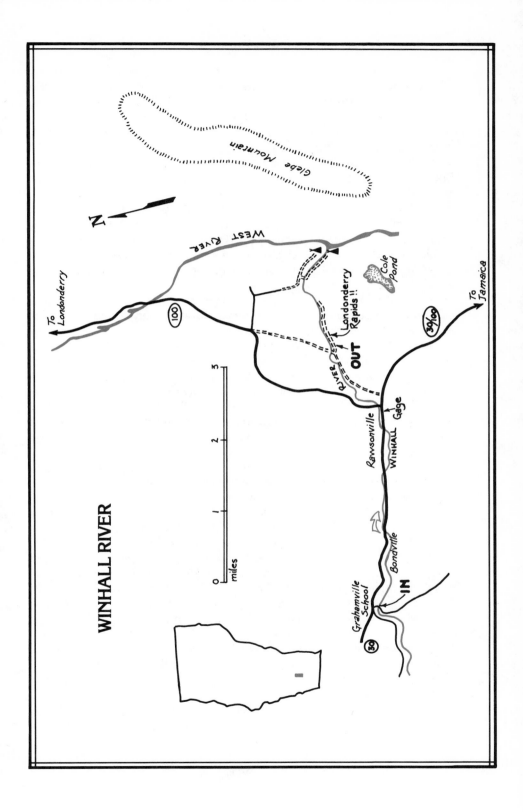

WINHALL RIVER

A suggested starting spot is near a small bridge about three miles west of Rawsonville on Route 30. Here the Winhall is about thirty feet wide with a scattering of small rocks. Just below, the river has gentle curves and easy standing waves as the paddler looks directly into a morning sun. In early spring this section, as others, has ice shelves on the banks extending into the river. In the beginning of a left turn, there's a barn on the left bank, a rock in the middle, and a 1-foot drop. Go to either side, although the right is a little trickier. Standing waves follow, then shortly, there's another rock in the right center and a 1- to 2-foot drop into more turbulence. Typical Class 2-3 rapids follow.

By a wooden store up on the left bank, the Winhall turns left in a fast, narrow chute blocked by a large rock in the middle. The left channel is S-shaped and the outflow is turbulent for 100 yards downstream, where under a small bridge there are several good hydraulics. This section is hard Class 2 or easy Class 3 at low water. The rapids taper down below the bridge, and houses follow the river for a while.

To the next bridge and the one beyond (green), the rapids are quite similar to those already seen. There are several sharp curves, and one section after a left turn has heavier-than-average haystacks (one to two feet at a gage reading of 1). Just as you glimpse houses in Rawsonville, a left turn has a sloping 3-foot drop with a big haystack at the bottom. Run it dead center.

The Route 100 bridge is next, and in the first left turn below it, a rock ledge pushes out from the right bank, creating a drop. It is easiest on the left or center (inclined one to two feet) and hardest on the extreme right (abrupt two feet). A small stream enters below from the left, and the Winhall then turns right. The rest of the way to the take-out is Class 2-3, depending on water level. The river here is shallower and wider than above. The take-out is at the next bridge on the Londonderry Road (dirt). Care should be taken not to proceed too far downstream, since the next left turn has a most difficult section — Londonderry Rapids (Class 4 even at low water).

The left turn that starts Londonderry Rapids is sharp, with a steep right bank and a beginning almost completely blocked by a huge, slab-like rock in the middle. To drop directly over this rock would mean a 4-foot fall. The approach must be made from either extreme. The right channel angles sharply right to left, then straightens out for 100 feet in some easier water. After this distance, rocks again choke the river, and a right or left route is necessary. The left is more technical while the right side is narrow, with the water falling over a sloping ledge. If this channel is chosen, you must keep your bow pointed downstream and blast through a powerful souse hole. This rapids should obviously be scouted before running. It should probably not be attempted when the gage reading is much less than 1.5, since there's not enough water. Most people — especially open boaters — should walk around if they plan to continue the trip down to the West.

There is a hand-painted gage on a rock on the left, upstream side, five feet from the Route 100 bridge in Rawsonville. If, during the spring runoff, the

Ball Mountain Dam is discharging 2,500 CFS or more, it is probably a good bet that the Winhall is runnable. This is certainly not always true, but it may save a long trip for nothing. It is estimated that a gage reading of 2.5 should make an interesting closed boat trip.

The Wardsboro is a nearby cousin of the Winhall. It is of similar character, but a little more difficult at comparable water levels.

Appendix I

Glossary

ACTIVE FACE — When considering a paddle blade it is the face which is pushing against the water.

BACKPADDLE — Paddling backwards to slow or check forward motion. Performed most vigorously when you're about to go over a waterfall.

BANANA BOAT — A C-1 or C-2.

BEAM — Widest part of boat.

BILGE — The point of greatest curvature in the bottom of a boat.

BLADE — The flat portion of the paddle that usually moves through the water.

BOW — Front of boat; usually points downstream.

BRACE — A sometimes static, sometimes dynamic stroke used to stabilize a boat. It acts similar to an outrigger. A brace can also be high or low. In a high brace, the hand grasping the paddle grip is much higher in the air than the hand on the throat. High braces (Duffek) are more frequently used in kayaks. In a low brace the paddle is close to horizontal. One "throws" or "hangs" a brace.

BROACHING — The act of having the canoe caught sideways to the current, particularly acute if a rock is on the downstream side. It's a potentially unstable situation, usually resulting in an upstream dump. After a broach and dump near a rock, the canoe usually assumes a classic pose wrapped around that rock, even if it's the only one for miles.

C-1	A covered canoe in which the paddler kneels and uses a single-bladed paddle. In the past, points on both bow and stern deck had to be higher than the middle, but this is no longer required. C-1s are generally made of fiberglass and weigh 25 to 50 lbs. They have more volume than kayaks and, hence, ride higher in the water.
C-2	A closed canoe designed for two persons.
CFS	Cubic feet per second — measure of water discharges (1 cubic foot of water equals about 8 gallons or 64 pounds).
CAPSIZE	Same as swamp. Your canoe is full of water and your pants are wet.
CHANNEL	The course a significant portion of water follows.
CHUTE	An inclined channel, usually narrower than the riverbed, down which water flows.
CLOSED BOAT	A term that encompasses all fiberglass or plastic boats where the deck is an integral part and not detachable, i.e., C-1, C-2, K-1.
COAMING	The rim-like area of a closed boat's cockpit under and around which the paddler attaches his skirt to the boat.
COCKPIT	In a closed boat, the area where a paddler sits or kneels.
CREST	The top of a standing wave.
CROSSCURRENT	A current moving at an angle to the main current. It's possible for a crosscurrent to be stronger than the main one.
CROSS DRAW	A stroke used to move the canoe, or part of the canoe, to the paddler's off side. The canoeist's hands do not change position on the paddle, although the paddle is crossed over the boat and placed in the water. It is used mainly by the bow in a C-2 or open boat and is not applicable in a K-1.
DECK	Term describing the covering which acts as a splash cover, placed over an otherwise open canoe. Also, describes the top half of a closed boat.
DISCHARGE	The volume of water that passes a specific cross-sectional area in a time interval — usually measured in cubic feet per second.

DOWNRIVER RACE	A long-distance race through white water against time. Also called "wildwater racing." Usually takes place over several miles and in difficult water.
DRAFT	The depth of water a boat displaces. Draw is also used in this sense, as, "My C-1 only draws three inches of water."
DRAW	Stroke used to move the canoe, or part of the canoe, towards the paddling side.
DROP	A sudden or abrupt increase in the river's gradient. The exact meaning depends on who is using the word. It is sometimes used to describe a set of rapids.
DUFFEL	Equipment of any kind carried in a canoe; usually refers to camping gear.
DUFFEK STROKE	Stroke named after the Czech paddler, Miko Duffek — used mainly by K-1s to enter or leave an eddy. To use it, the paddler leans his boat into the eddy (when entering) and rests his weight on the paddle, which is almost vertical. It is executed as a dynamic maneuver. When used in a C-1, it is sometimes called a "hanging draw." If the stroke isn't dynamic, you're wet again.
DUMP	The overturning of a boat with possible eviction of the boater.
EDDY	A relatively calm area of water usually found downstream from rocks and near the shore line. Once thought to be dangerous, they are now routinely used for rest stops. The current in an eddy is usually moving upstream.
EDDY HOPPING	The act of using eddies to move either downstream or upstream. Very effective when paddling an unfamiliar river and for scouting from the boat.
EDDY LINE	The line, or thin area, dividing the upstream eddy current and the downstream current of the river. Many dumps occur here. Also called a differential or boundary line.
EDDY TURN	A dynamic technique used by boaters to enter an eddy. Also referred to as eddying out.
END OVER END (ENDER)	A rather spectacular maneuver whereby a closed boat is driven upstream into the trough of a wave, is tilted up until vertical and then completely over. Can be done forwards or backwards.

ENGLISH GATE	A single slalom gate that can be practiced in still or moving water. The paddler passes through the gate in a variety of ways following a standard pattern. The course requires both forward and backward paddling and four rolls. It is used to improve slalom skills.
EXPOSED ROCK	A rock without its raincoat.
FACE	Name given to the two sides of the blade.
FALLS	Term usually used to describe a waterfall. In some areas it also means rapids, or hard rapids. If use is unclear, assume the worst.
FEATHER	In the recovery stroke, the act of pivoting the paddle 90° about its long axis to offer minimum resistance to wind or water. Also refers to the fact that the blades of a kayak paddle are rotated 90° from each other.
FERRY	Act of moving the boat sideways without progressing upstream or downstream. If the bow is pointed down-stream, the word "setting" is often used.
FIXED HAND	On a kayak paddle, the hand which does not change its grip upon feathering. It controls the angle of either blade.
FLOTATION	Material placed in a canoe to keep it from sinking during a swamp. Air bags or foam-type materials are quite common. Very necessary if you want a lasting relationship with your boat.
FOLDBOAT	A canoe whose outer canvas skin is supported by an ingenious collapsible wood frame. A popular craft of years past. Great boat for carrying in elevators.
GAGE	A device used to measure the level of the river. Also spelled "gauge."
GRAB LOOPS	A closed loop of rope on the bow or stern of canoes and kayaks used for holding on to the boat after a dump.
GRADIENT	The decrease in elevation of a river as a function of dis-tance. Measured in feet per mile.
GUNWALE	The upper edge on the side of an open canoe. The "w" is silent for some reason.
HAIR	Term sometimes used to describe heavy water.
HAYSTACK	See *Standing Wave*.

HEAVY WATER	Rapids where the standing waves, souse holes, and hydraulics are particularly big and powerful. Discharge is usually high.
HULL	The lower half of a closed boat. Also a southern "hole."
HYDRAULIC	Formally known as a dynamic hydraulic jump, it is a water formation caused by a sudden drop of the riverbed over a dam, rock, or ledge, causing an upstream current to be formed at the surface at the base of the drop, with an associated downstream current found deeper. Water is usually aerated and offers poor support for a boat. A hydraulic can be dangerous with its holding power.
INVERTED J	Guiding stroke, sometimes used in solo canoeing when backpaddling. Stroke is similar to a "normal J," only it is executed in a reverse direction.
J STROKE	Stroke used by a stern or solo paddler to correct the characteristic swing to the off side when a forward stroke is taken. It is so called because the paddle follows the shape of a J. This stroke is sometimes converted to an L or a C.
K-1	A closed boat (kayak) derived from the Eskimos where the paddler sits with legs extended forward and uses a double-bladed paddle.
KEEL	A projection from the bottom-most part of the hull extending from bow to stern. Used to help the paddler keep a straight line or keep from being blown sideways by wind. They are used mostly on lake canoes. White water canoes have little or no keel.
KEEPER	A souse hole or hydraulic that is powerful enough to hold a boat or boater for an extended time.
KNEE BRACES	Supports that fit over the knee area and are attached to the boat. Used to strengthen the contact of boater to boat. They help to give better body boat control.
LEAN	The act of leaning the canoe for the purpose of stabilizing or maneuvering the boat — very important in closed boats.
LEDGE	A projecting ridge of rock over which the river drops either gradually or abruptly.

LINING	The act of using a painter to guide the boat through a rapids or other obstacle from shore. Upstream lining is called "tracking."
MACHINE	Rigid supports placed inside a closed boat (usually C-1 or C-2) that keep a boater's legs from moving about, thus, giving more body control of the boat. Term originated with the German boater, Wolfgang Peters.
MEANDER	Term used to describe the bends or turns of a river. Word originated from a wandering Turkish river known as the Menderes.
MOUTH	Area where a river joins another body of water.
OFF SIDE	Side of the canoe on which a paddler does not have his paddle; very vulnerable to dumps in a C-1.
OPEN CANOE	A canoe as usually envisaged in a word assocation test. The boat has no permanent deck.
OUTSIDE BANK	In a curve or turn, the side toward which centifugal force pushes the current.
OXBOW	The loop in a river that has ceased to be a main channel due to the silting up of the ends after the river has cut through the land within the meander at a narrower place. Also, a ribbon placed around an ox's head.
PADDLE	The tool, when properly used, that propels the boat. Paddles come in many shapes, sizes, and materials.
PAINTER	The thin lines attached to either end of a canoe, generally used for lining and rescue.
PILLOW	A wave protecting or buffeting the upsteam side of a rock or other solid object.
PITCH	Term sometimes used to descirbe a heavy drop, rapids, or waterfall.
PLAYING	The practicing of a paddling technique — usually in a localized area like a souse hole, a sand box, or rapids. Frequently, a paddler consciously places himself in a situation that an inexperienced person would try to avoid. Great for improving skill.
POLING	Technique whereby a long pole is used to guide the canoe, instead of a paddle. Can be used to take a canoe

either up or downstream. Not used frequently in white water.

POOL — An area of quiet water with little or no current.

POP UP — Almost an ender where the hole or wave trough doesn't flip the boat, but merely pops it back (usually downstream).

PORTAGE — Term describing delightful act whereby a canoeist carries his boat around some obstacle. Also called a carry.

POWER FACE — Face of paddle which is actually pushing up against the water.

PRY — Stroke used to move boat away from paddling side while paddle is still on the paddling side. The canoe is frequently used as a fulcrum.

PUT-IN — The start of a trip. This word is used as both a noun and a verb.

RAPIDS — In general, that portion of a river where the water is quick-moving and the surface is broken by obstacles, forms a series of waves, or both. This word is used as both singular and plural. When describing a specific piece of white water, common useage is singular.

RATING TABLE — A relationship correlating the gage height with the actual amount of water flowing for a particular river.

READING — Skill consisting of ascertaining water formations, difficulties, and possible routes. A canoeist "reads" water.

RECOVERY STROKE — Act of getting ready for the next stroke; not really a stroke. Can be done with paddle in or out of water. Blade is usually feathered.

RIBS — Reinforcements (used for strengthening a hull or deck) that usually run perpendicular to the keel direction.

RIP — Term used to describe a rapids.

ROCK GARDEN — Rapids generously supplied with rocks.

ROCKER — The curved keel line in the hull of a boat. When viewed sideways, a C-1's rocker gives it a banana shape. In general, the more rocker, the easier a boat is to pivot.

ROLLING — Maneuver whereby a closed boat is righted from an upside-down position by the paddler. Also called, but not often, an esquimantage.

RUDDER — Position whereby a paddler holds his paddle in the water at a fixed angle for the purpose of steering. Boat must be moving faster than current for it to be effective.

RUN — Word used to describe a particular river trip; also, the act of making a trip.

SAULT — An esoteric word for rapids.

S-TURN — A meander shaped like an S. A sharp S-turn is a Z.

SCOUT — Looking over rapids or difficult spots before running; it can be done from boat or shore.

SCULLING — Alternately drawing and prying a paddle in a continuous figure-8 pattern for the purpose of fine alignment of a canoe. Can be used to move canoe towards or away from paddling side, and can be used in conjunction with another stroke.

SELF RESCUE — Act of rescuing oneself when in trouble. Very important as others can only help most effectively when the action stops.

SHAFT — Part of paddle between handle and blade.

SHOE KEEL — A keel having a minimum extension. Usually used on aluminum white water boats.

SHOCK CORD — Elastic cord of various diameters used in securing paddles and other equipment.

SHUTTLE — The placing of cars at the start and take-out spots which allows paddlers to get back to civilization.

SIDESLIPPING — Situation where a canoe's center of gravity continues on in the initial direction of movement, even though the boat may be turning, e.g., a canoe will continue downstream a short way during the execution of an eddy turn. A similar effect is seen in skiing.

SKIRT — Garment worn around the waist of closed boat paddlers that attaches to the coaming to make the cockpit water-tight. (It is not for the purpose of telling boy from girl paddlers.)

SLALOM — A race against time in which a paddler must negotiate a specified course designated by pairs of vertically suspended poles (gate). The poles are colored either red and white, green and white, or black and white, and the paddler must pass through a gate always keeping the green and white pole on the right. Some gates are designed to be gone through backwards and others in an upstream direction. Touching a gate results in adding penalty times to the final score. The classes of a slalom are determined by boat type and sex of paddlers. Lowest score wins.

SLICE
Word used to describe the stroke where the edge of the paddle acts as the power face. It is used to move the paddle through the water with a minimum of friction, e.g., when recovering, without taking the blade out of the water.

SOLOING
Paddling a boat single, usually restricted to open boats or C-2s.

SOUSE HOLE
A water formation in which there is actually a hole surrounded by water. It can be formed in a variety of ways and can be used for playing or for getting into trouble.

SPOON
Term describing the curved shape of some kayak paddle blades.

STANDING WAVE
Similar in geometry to ocean waves except it does not propagate in any direction, although it can drastically change its shape with time. They usually appear in a series, and are one way a river dissipates energy. Also called haystack.

STERN
Rear of boat — usually points upstream. What the Coast Guard is if you're caught without a lifejacket on a river.

STOPPER WAVE
A wave that usually follows an abrupt drop or souse hole; it rises sharply, tending to stop a boat and tilt it vertically. Usual strategy is to avoid them or blast through.

SURFING
Act of riding the upstream side of a wave. Since the current is trying to force the boat up the wave (and gravity down), a dynamic equilibrium may be achieved whereby a boater, by using his paddle for braces, can remain for an extended period.

SWAMP
Act of canoe filling up with water — a most exasperating situation.

SWEEP
A wide, shallow stroke used for turning the canoe away from the paddle side. A reverse sweep does the opposite. Also, the last boat in a group.

T-GRIP
The shape of many paddle handles. It gives better paddle

control than traditional types. You will frequently be able to identify your paddle by the finger marks imbedded in the T-grip.

TAKE-OUT — The opposite of put-in.

TECHNICAL — Word used to describe a rapid where there are many intricate passageways and much maneuvering.

THIGH STRAPS — Straps used for same purpose as machines.

TRACK — Paddling in a straight line.

THROAT — Part of paddle immediately above blade.

THROW LINE — A length of rope used for throwing to (not at) a troubled boater in a rescue attempt.

THWART — The cross supports that run from gunwale to gunwale in an open boat.

TOE BLOCKS — Supports against which a closed boater braces his feet — used in conjunction with thigh straps or machines.

TRIM — Term relating to the way a boat sits in the water, usually when carrying a load. Also, what all boaters should be.

TROUGH — The low point just before and after the crest of a standing wave.

UPSTREAM SIDE — When a canoe is perpendicular to the current, the side of the boat closest to the upstream side. It is also the side to which most canoes dump.

WATERSHED — Land area that is drained by a river.

WET PACK — A small, rubber waterproof pack that folds in such a way as to be watertight. It is used for carrying lunches, etc.

WHITE WATER — Water that has been aerated — and which will cause foaming at the mouth of the most ardent enthusiasts.

Appendix II

For those few times when an emergency happens and you need medical attention in a hurry, the following list is provided to indicate the closest hospital. This list should be consulted before you start a trip so you are prepared and don't waste valuable time in a crisis. Even though this list was compiled with care, its accuracy cannot be guaranteed in all cases. Even if the phone numbers have changed, at least you know what town and hospital to ask a telephone operator for if you want to call.

RIVER	TOWN OF HOSPITAL	NAME, ADDRESS & PHONE NUMBER
Ammonoosuc (NH)	Littleton	Littleton Hospital 107 Cottage St. 603-444-7731
	Lancaster	Beatrice D. Weeks Memorial Middle St. 603-788-4911
Androscoggin (NH)	Berlin	Androscoggin Valley Hospital 324 School St. 603-752-2200
Ashuelot (NH)	Keene	Cheshire Hospital 580 Court St. 603-352-4111
Bantam (CT)	Torrington	Charlotte Hungerford Hospital 540 Litchfield St. 203-482-9351
Bearcamp (NH)	North Conway	Memorial Hospital Intervale Rd. 603-356-5461
	Wolfboro	Cottage Hospital Swiftwater Rd. 603-747-2761

Black (VT)	Windsor	Mount Ascutney Hospital Country Rd. 802-674-6711
	Springfield	Springfield Hospital 25 Ridgewood Rd. 802-885-2151
Blackledge (CT)	Willimantic	Windham Community Memorial Hospital 112 Mansfield Ave. 203-423-9201
	Norwich	Norwich Hospital Route 12 203-852-2000
		William W. Backus Hospital 326 Washington St. 203-889-8331
Blackwater (NH)	Concord	Concord Hospital 250 Pleasant St. 603-225-2711
		New Hampshire Hospital 105 Pleasant St. 603-224-6531
Boreas (NY)	Ticonderoga	Moses Lundington Hospital Montcalm and Wicker St. 518-585-2831
	Glens Falls	Glens Falls Hospital 190 Park St. 518-792-3151
Chickley (MA)	Greenfield	Franklin County Public Hospital 164 High St. 413-772-0211
	North Adams	North Adams Regional Hospital Hospital Ave. 413-663-3701
Cold (NH)	Bellows Falls (VT)	Rockingham Memorial Hospital Hospital Center 802-463-3903
Contoocook (NH)	Concord	Concord Hospital 250 Pleasant St. 603-225-2711

	Concord	New Hampshire Hospital 105 Pleasant St. 603-224-6531
	Peterborough	Monadnock Community Hospital Old Street Rd. 603-924-7191
Dead (ME)	Jackman	Marie Joseph Hospital Main St. 207-668-2691
	Skowhegan	Redington-Fairview General Hospital Fairview Ave. 207-474-5121
Deerfield (MA)	Greenfield	Franklin County Public Hospital 164 High St. 413-772-0211
	North Adams	North Adams Regional Hospital Hospital Ave. 413-663-3701
Ellis (NH)	North Conway	Memorial Hospital Intervale Rd. 603-356-5461
Farmington (CT)	Winsted	Winsted Memorial Hospital 115 Spencer St. 203-379-3351
	Hartford	Hartford Hospital 80 Seymour St. 203-524-3011
		Mount Sinai Hospital 500 Blue Hills Ave. 203-242-4431
		St. Francis Hospital 114 Woodland St. 203-548-4000
Gale (NH)	Littleton	Littleton Hospital 107 Cottage St. 603-444-7731
Green (MA)	Greenfield	Franklin County Public Hospital 164 High St. 413-772-0211

Housatonic (CT)	New Milford	New Milford Hospital 21 Elm St. 203-354-5531
Hudson (NY)	Ticonderoga	Moses Ludington Hospital Montcalm and Wicker St. 518-585-2831
	Glens Falls	Glens Falls Hospital 100 Park St. 518-792-3151
Indian (NY)	Ticonderoga	Moses Ludington Hospital Montcalm and Wicker St. 518-585-2831
	Glens Falls	Glens Falls Hospital 100 Park St. 518-792-3151
Jeremy (CT)	Willimantic	Windham Community Memorial Hospital 112 Mansfield Ave. 203-423-9201
Mad (NH)	Plymouth	Sceva Speare Memorial Hospital Hospital Rd. 603-536-1120
Mascoma (NH)	Lebanon	Alice Peck Day Memorial Hospital 125 Mascoma St. 603-448-3121
	Hanover	Mary Hitchcock Memorial Hospital 2 Maynard St. 603-643-4000
Millers (MA)	Athol	Athol Memorial Hospital 2033 Main St. 617-249-3511
	Montague	Farren Memorial Hospital 56 Main St. 413-774-3111
North (MA)	Greenfield	Franklin County Public Hospital 164 High St. 413-772-0211
Ompompanoosuc (VT)	White River Junction	Veterans Administration Center N. Hartland Rd. 802-295-9363

	Randolph	Gifford Memorial Hospital 44 S. Main St. 802-728-3366
Otter Brook (NH)	Keene	Cheshire Hospital 580 Court St. 603-352-4111
Pemigewasett (NH)	Littleton	Littleton Hospital 107 Cottage St. 603-444-7731
	North Conway	Memorial Hospital Intervale Rd. 603-356-5461
	Plymouth	Sceva Speare Hospital Hospital Rd. 603-536-1120
Piscataquog (NH)	Manchester	Catholic Medical Center 200 Hanover St. 603-668-3545
		Elliot Hospital 955 Auburn St. 603-669-5300
		Veterans Administration Hospital 718 Smyth Rd. 603-624-4366
Quaboag (MA)	Palmer	Wing Memorial Hospital Wright St. 413-283-7651
	Ware	Mary Lane Hospital 85 South St. 413-967-6211
Rapid (ME)	Berlin (NH)	Androscoggin Valley Hospital 324 School St. 603-752-2200
	Colebrook (NH)	Upper Connecticut Valley Hospital Hospital Rd. 603-237-4971
Sacandaga (NY)	Gloversville	Nathan Littauer Hospital 99 E. State St. 518-725-8621

	Amsterdam	Amsterdam Memorial Hospital Upper Market St. 518-842-3100
		St. Mary's Hospital 427 Guy Park Ave. 518-842-1900
Saco (NH)	North Conway	Memorial Hospital Intervale Rd. 603-356-5461
Salmon (CT)	Willimantic	Windham Community Memorial Hospital 112 Mansfield Ave. 203-423-9201
Sandy (MA)	Winsted (CT)	Winsted Memorial Hospital 115 Spencer St. 203-379-3351
Saxtons (VT)	Bellows Falls	Rockingham Memorial Hospital Hospital Ct. 802-463-3903
Shepaug (CT)	New Milford	New Milford Hospital 21 Elm St. 203-354-5531
Smith (NH)	Franklin	Franklin Regional Hospital Aiken Ave. 603-934-2060
	Plymouth	Sceva Speare Memorial Hospital Hospital Rd. 603-536-1120
Souhegan (NH)	Peterborough	Monadnock Community Hospital Old Street Rd. 603-924-7191
	Nashua	Nashua Hospital 8 Prospect St. 603-883-5521
		St. Joseph Hospital 172 Kinsley St. 603-889-6681
Sugar (NH)	Claremont	Claremont General Hospital 243 Elm St. 603-542-7771

	Newport	Newport Hospital 167 Summer St. 603-863-1123
Stony Brook (NH)	Peterborough	Monadnock Community Hospital Old Street Rd. 603-924-7191
	Nashua	Nashua Hospital 8 Prospect St. 603-883-5521
		St. Joseph Hospital 172 Kinsley St. 603-889-6681
Suncook (NH)	Concord	Concord Hospital 250 Pleasant St. 603-225-2711
		New Hampshire Hospital 105 Pleasant St. 603-224-6531
Swift (NH)	North Conway	Memorial Hospital Intervale Rd. 603-356-5461
Ten Mile (NY)	New Milford (CT)	New Milford Hospital 21 Elm St. 203-354-5531
Waits River (VT)	Woodsville (NH)	Cottage Hospital Swiftwater Rd. 603-747-2761
	Barre (VT) Berlin (VT)	Central Vermont Hospital 802-229-9121
Wardsboro River (VT)	Townshend	Grace Cottage Hospital Route 35 802-365-7920
West River (VT)	Townshend	Grace Cottage Hospital Route 35 802-365-7920
Westfield (MA)	Westfield	Noble Hospital 115 W. Silver St. 413-568-2811
		Western Massachusetts Hospital 91 E. Mountain Rd. 413-562-4131

	Northampton	Cooley Dickinson 30 Locust St. 413-584-4090
White (VT)	White River Junction	Veterans Administration Center N. Hartland Rd. 802-295-9363
	Randolph	Gifford Memorial Hospital 44 S. Main St. 802-728-3366
Wild Ammonoosuc (NH)	Woodsville	Cottage Hospital Swiftwater Rd. 603-747-2761
	Littleton	Littleton Hospital 107 Cottage St. 603-444-7731
Williams(VT)	Springfield	Springfield Hospital 25 Ridgewood Rd. 802-885-2151
Winhall (VT)	Townshend	Grace Cottage Hospital Route 35 802-365-7920

Appendix III

OTHER GUIDEBOOKS TO NEW ENGLAND WATERS

AMC River Guide I: Maine. Appalachian Mountain Club, 5 Joy St.,
  Boston, MA (1980).
AMC River Guide II: Central and Southern New England. Appalachian
  Mountain Club, 5 Joy St., Boston, MA (1978).
No Horns Blowing. Eben Thomas.
  Hallowell Printing Co., Hallowell, ME (1973).
Hot Blood and Wet Paddles. Eben Thomas.
  Hallowell Printing Co., Hallowell, ME (1974).
Canoeing Maine #1 and #2. Eben Thomas.
  The Thorndike Press, Thorndike, ME (1979).
Canoe Camping. Roioli Schweiker.
  New Hampshire Publishing Co. (1977).
Canoeing Trips in Connecticut. Pamela Detels and Janet Harris.
  Pequot Press, Chester, CT (1977).
The Connecticut River Guide. The Connecticut River Watershed
  Council. 125 Combs Rd., Easthampton, MA 01027 (1974).
The Farmington River and Watershed. The Farmington River
  Watershed Association, Inc., Avon, CT (1970).
Charles River Canoe Guide. The Charles River Watershed Association,
  Auburndale, MA (1973).
Adirondack Canoe Waters. P.F. Jamieson.
  Adirondack Mountain Club, Inc., Glen Falls, NY (1975).
The Nashua River Canoe Guide. Charles Harris.
  Nashua River Watershed Association, Ayer, MA (1978).
The Saco River: A History and Canoeing Guide. Viola Sheehan.
  Saco River Corridor Association, Saco, ME.
Canoeing Massachusetts, Rhode Island and Connecticut.
  Ken Weber. New Hampshire Publishing Co. (1980).

Appendix IV

BOOKS ON CANOEING AND KAYAKING TECHNIQUES

A White Water Handbook for Canoe and Kayak, 2nd ed.
    John Urban and T. Walley Williams. Appalachian Mountain Club,
    5 Joy St., Boston, MA (1981).
Kayaking. Jay Evans and Robert Anderson.
    The Stephen Greene Press, Brattleboro, VT (1975).
Basic River Canoeing. Robert NcNair.
    Buck Ridge Ski Club, Swarthmore, PA (1968).
Recreational White Water Canoeing. Thomas Foster.
    Leisure Enterprise, 8 Pleasant St., Millers Falls, MA (1978).
The All Purpose Guide to Paddling. Dean Norman.
    Great Lakes Living Press, Matteson, IL (1976).
Boat Builders Manual. Charlie Walbridge, ed.
    Wildwater Designs, Penllyn, PA 19422 (1979).
The Complete Wilderness Paddler. J.W. Davidson and J. Rugge.
    Alfred A. Knopf, Inc., New York, NY (1976).
Canoeing. The American National Red Cross.
    Doubleday and Co., Garden City, NY (1977).
Running the Rivers of North America. Peter Wood.
    Barre Publishing, Barre, MA (1978).
Canoeing. Mike Michaelson and Keith Ray.
    Henry Regnery Co., Chicago, IL (1975).
Whitewater Rafting. William McGinis.
    Times Books, NY (1975).
Pole, Paddle, and Portage. Bill Riviere.
    Little, Brown and Co., Boston, MA (1969).
A Guide to Big Water Canoeing. David Herzog.
    Contemporary Books, Inc., Chicago, IL (1978).
Living Canoeing. Alan Byde.
    Adam and Charles Black, Ltd., London (1972).

You, Too, Can Canoe. John Foshee.
   Stride Publishers, Inc., Huntsville, AL (1977).
Building and Repairing Canoes and Kayaks. Jack Brosius and David LeRoy.
   Contemporary Books, Inc., Chicago, IL (1978).
*Wildwater: The Sierra Club Guide to Kayaking and White Water
   Boating.* Lito Tejada-Flores. Sierra Club Books,
   San Francisco, CA (1978).
The Canoe and White Water. C.E.S. Franks.
   Univ. of Toronto Press, Toronto (1977).
Wilderness Canoeing. John Malo.
   Collier Press (1971).

Appendix V

WHITE WATER CANOE CLUBS IN NEW ENGLAND

Appalachian Mountain Club, 5 Joy St., Boston, MA 02108.
  Chapters: *Boston, Berkshire, Connecticut, Delaware, Pennsylvania, Maine, New York, New Hampshire, S.E. Massachusetts, Worcester, Rhode Island.*
Dover Dunkers, QYB Cabin, West Dover, VT 05356
Experiment with Travel, 281 Franklin St., Box 2452, Springfield, MA 01101
Fairfield White Water Association, 100 Vermont Ave., Fairfield, CT 06430
Hampshire College Kayak Program, Amherst, MA 01002
Ledyard Canoe Club, PO Box 9, Hanover, NH 03755
Marlboro College Outdoor Program, Marlboro, VT 05344
MIT Outdoor Club, Cambridge, MA
Northern Vermont Canoe Cruise, Winooski, VT 05404
Penobscot Paddle and Chowder Society, Box 121, Stillwater, ME 16801
Kayak & Canoe Club of Boston, Boston, MA
Greenfield Community College Aquatics Program, Greenfield, MA

Appendix VI

Rivers proposed for study to become wild or scenic rivers under the Wild &
Scenic Rivers Act

MAINE

River	Section	Mileage
Allagash		
Aroostook	a) Washburn to Sheridan	17
	b) Confluence with Machias to Millinocket Stream	42
Baker Branch	St. John to headwaters	46
Crooked	Sebago Lake to headwaters	45
Kennebec	Bay Point to Bath	11
Kirby Stream	Forks to headwaters of Kirby Stream	42
Spencer Stream and *Dead* (including)		
Kennebec	Wyman Lake to Harris Dam	25
South Branch: Dead	Flagstaff Lake to headwaters	25
Machias	a) Whitneyville Reservoir to Crooked River	30
	b) Aroostook River to Big Machias Lake	32
Mattawamkeag (including)	Mattawamkeag to Haynesville	45.5
Macwahoc Stream	Confluence with Mattawamkeag to headwaters	
Moose	Attean Pond to Canada	39
Narraguagus (including)	Milbridge to East Branch headwaters	54
Schoodic Branch	Confluence with Narraguagus River to Route 193	15

East Branch: Nezinscot	Buckfield to headwaters	17
Passadumkeag (including)	Passadumkeag to headwaters	41
Cold Stream	Passadumkeag to Enfield	4
Little Cold Stream	Cold Stream to headwaters	3
Penobscot	Howland to Mattawamkeag	22
East Branch	Medway to Whetstone Falls	17
	Whetstone Falls to Grand Lake Matagamon	
North Branch	Pittston Farm to headwaters	27
West Branch	Chesuncook Lake to Seboomook Lake	25
Pleasant	Columbia Falls to Pleasant River Lake	30
Seboeis	East Branch: Penobscot to Snoeshoe Lake	27
Saco	a) E. Hiras to E. Brownfield	13
	b) E. Brownfield to Fryeburg	22
Sheepscot (including)	N. of Wiscasset to Somerville	30
Marsh	Confluence with Sheepscot to New Castle	6.5
Dyer	Confluence with Sheepscot to North New Castle	4
West Branch: Sheepscot	a) Confluence with Sheepscot to SW of Windsorville	6
	b) Somerville to headwaters	20
St. Croix	Grand Falls to Vanceboro	28
St. Francis	Estcourt to confluence with St. John	56
St. John	Dickey to confluence with Baker Branch	76.5
(including)		
Allagash	Confluence with St. John to Twin Brook Rapids	6
Big Block	Confluence with St. John to Canadian border	29
Little Block	Confluence with St. John to headwaters	27
Southwest Branch	Baker Branch confluence to Little St. John Lake	35

NEW HAMPSHIRE

Androscoggin	Pontook Reservoir to Errol	13
Connecticut	a) Dalton to N. Stratford	40
	b) N. Stratford to Beecher Falls	22

Dead Diamond	Confluence with Magalloway to headwaters	12
(including)		
Magalloway	Below Aziscohos Dam to Umbagog Lake	16
Nash Stream	Confluence with Ammonoosuc to headwaters	14
East Branch:		
Pemigewasset	Boyle Brook to headwaters of East Branch	16
(including)		
N. Fork of East Branch	Confluence with East Branch to headwaters	8
Franconia Branch	Confluence with East Branch to headwaters	6
Perry Stream	Confluence with Lake Francis to headwaters	18
Phillips Brook	Confluence with Ammonoosuc to headwaters	16
Saco	N. Conway to headwaters	27
Swift Diamond	Confluence with Dead Diamond to headwaters	18

VERMONT

Batten Kill	Route 22 to near Arlington	18
Deerfield	N. of Searsbury Reservoir to headwaters	11
Missisquoi	Canadian border to headwaters of Burgess Branch	25
North Branch:		
Nulhegan	Connecticut River to headwaters of North Branch	20
(including)		
East Branch Nulhegan	Nulhegan River to near Little Averill Lake	12
Ottauquechee	Woodstock to headwaters	23
West	Brattleboro to Townshend Dam	19
	Ball Mountain to headwaters	22
White	a) Confluence with Connecticut to S. Royalton	18
	b) S. Royalton to Rochester	21
Winhall	Bondville to headwaters of East and West Branches	10.5

APPENDIX VII
RATING TABLES
GAGE READING VS. CFS DISCHARGE

River	Gage Location	1.0	1.1	1.2	1.3	1.4	1.5	1.6
Ammonoosuc	Bath	65	76	87	99	113	128	146
	Bethlehem	30	38	47	57	68	80	92
Androscoggin	Errol	485	537	591	652	716	780	860
Assabet	Maynard	1.0	2.2	3.95	6.18	8.93	12.2	16.1
Ashuelot	Gilsum Gorge	8	10	12	14	17	21	24
	Hinsdale							
S. Br. Ashuelot	Webb (Rte. 12)							
Blackwater	Webster							
Branch	Forestdale							
Cold	Drewsville							
Contoocook	Peterborough							
	Henniker							
Deerfield	Charlemont							
Ellis	Jackson (Rte. 16)	9	13	18	25	32	43	54
Farmington	New Boston							
	Tariffville	220	275	330	390	450	510	575
Green	E. Colrain							
Housatonic	Gaylordsville							
Hudson	N. Creek							
Mascoma	Mascoma (Rte. 4A)	68	83	99	116	135	155	175
Millers	S. Royalston							
	Farley							
North	Shattuckville							
Piscataquog	Grasmere							
Quaboag	W. Brimfield							
Salmon	Comstock Bridge							
Sandy	Robertsville							
Saxtons	Saxtons River							
Sacandaga	Griffin							
	Hope							
Smith	Bristol							
Sugar	W. Claremont	41	58	76	97	123	150	181
Ten Mile	Webatuck	56	68	81	96	115	135	155
West	Jamaica							
Westfield	Knightville							
	Huntington						123	150
White	W. Hartford							
Williams	Brockway Mills						51	63

Gage reading (feet)

1.7	1.8	1.9	2.0	2.1	2.2	2.3	2.4	2.5	2.6	2.7
167	191	219	249	279	310	342	374	407	442	478
105	120	136	153	171	190	210	231	253	278	305
935	1022	1110	1203	1301	1406	1515	1634	1760	1892	2023
21.1	27.1	34.3	43.1	53.5	65.4	79.6	96.2	115	135	157
28	33	37	42	47	53	59	65	72	79	86
			23	30	37	46	56	67	80	94
			33	43	53	65	79	93	108	125
			120	139	160	184	210	238	268	300
66	79	82	109	126	145	165	187	210	235	262
640	705	770	840	910	980	1060	1140	1220	1300	1400
								49	60	73
			450	490	535	580	630	680	740	800
			170	199	230	265	304	346	392	442
200	230	260	293	328	365	405	448	493	540	590
								229	270	315
			52	63	75	88	103	119	136	154
				9.5	15	23	31	41	52	65
			107	122	140	160	180	200	225	250
			72	82	93	105	117	131	147	165
			295	355	414	480	552	630	715	804
			36	42	48	54	61	69	79	90
215	252	292	335	380	429	480	533	589	648	709
175	200	225	255	285	315	345	380	415	450	485
179	211	247	286	330	374	421	471	525	583	645
77	93	109	126	145	164	185	207	231	256	282

APPENDIX VII cont'd

RATING TABLES
GAGE READING VS. CFS DISCHARGE

River	Gage Location	2.8	2.9	3.0	3.1	3.2	3.3	3.4
Ammonoosuc	Bath	515	555	600	645	695	745	795
	Bethlehem	335	365	400	435	470	505	540
Androscoggin	Errol	2151	2284	2420	2555	2693	2834	2978
Assabet	Maynard	180	203	228	254	281	309	339
Ashuelot	Gilsum Gorge	94	102	111	121	131	142	154
	Hinsdale	42	54	67	84	102	124	148
S. Br. Ashuelot	Webb (Rte. 12)			39	44	48	53	58
Blackwater	Webster	110	125	141	159	177	197	219
Branch	Forestdale	142	161	182	205	229	255	283
Cold	Drewsville			103	116	129	142	156
Contoocook	Peterborough Henniker	335	371	410	451	495	541	550
Deerfield	Charlemont			911				
Ellis	Jackson (Rte. 16)	291	321	353	388	425	463	504
Farmington	New Boston			99	115	134	154	178
	Tariffville	1500	1600	1700	1800	1900	2000	2100
Green	E. Colrain	88	105	125	147	171	197	225
Housatonic	Gaylordsville	860	930	1000	1080	1160	1240	1330
Hudson	N. Creek	497	555	619	687	760	839	922
Mascoma	Mascoma (Rte. 4A)	643	699	758	819	883	950	1021
Millers	S. Royalston Farley	365	419	480	538	601	668	740
North	Shattuckville	173	194	217	241	267	295	324
Piscataquog	Grasmere			16	22	30	40	51
Quaboag	W. Brimfield	79	94	111	129	148	169	191
Salmon	Comstock Bridge							
Sandy	Robertsville	275	300	330	360	390	420	450
Saxtons	Saxtons River			29	36	46	57	68
Sacandaga	Griffin	184	204	224	244	265	287	310
	Hope	899	1001	1110	1220	1337	1460	1593
Smith	Bristol	102	116	130	147	165	185	207
Sugar	W. Claremont	773	840	907	977	1049	1124	1201
Ten Mile	Webatuck	520	560	600	640	680	730	780
West	Jamaica							
Westfield	Knightville			134	158	184	212	241
	Huntington	715	790	870	954	1040	1140	1240
White	W. Hartford			197	229	264	302	343
Williams	Brockway Mills	310	340	368	397	428	460	495

Gage reading (feet)

3.5	3.6	3.7	3.8	3.9	4.0	4.1	4.2	4.3	4.4	4.5
845	900	955	1010	1070	1130	1190	1260	1330	1400	1470
575	610	645	680	715	755	800	845	895	945	995
3120	3266	3414	3565	3709	3855	4003	4153	4305	4455	4606
370	402	435	469	505	543	582	623	666	710	756
166	178	191	204	218	232	248	265	283	301	320
174	203	236	272	317	366	420	480	547	620	647
63	69	78	87	100	118	143	176	215	252	293
241	266	292	320	349	380	412	445	479	515	552
312	343	375	408	442	478	516	555	592	629	668
171	188	207	227	248	270	295	322	351	381	413
636	684	734	786	840	876	954	1014	1076	1140	1202
1360					1870					2460
547	586	627	669	713	758	805	854	904	957	1010
205	234	268	305	345	384	427	472	519	570	620
2200	2320	2440	2560	2680	2800	2970	3040	3160	3280	3400
255	292	333	376	424	475	525	578	635	694	757
1420	1510	1610	1710	1810	1920	2030	2140	2250	2370	2490
1011	1106	1206	1313	1427	1547	1673	1807	1947	2095	2251
1094	1171	1250	1344	1441	1544					2072
					65	80	98	119	143	169
811	887	967	1050	1140	1230	1320	1420	1520	1620	1720
355	390	426	465	509	555	606	660	717	777	840
63	77	93	111	131	153	177	204	233	265	299
215	239	264	291	319	349	380	412	445	480	516
					180	215	255	300	350	400
480	520	560	610	660	710	760	810	860	920	980
81	96	112	130	149	170	193	218	245	275	306
334	358	383	410	437	466	495	525	556	588	621
1733	1880	2027	2180	2340	2501	2667	2840	3020	3207	3400
230	257	287	318	353	390	430	473	518	568	620
1280	1362	1445	1532	1620	1711	1804	1900	1997	2098	2200
830	880	930	990	1050	1110	1170	1230	1300	1370	1440
					34	46	59	76	93	113
272	305	339	374	411	450	492	535	583	632	684
1340	1440	1540	1640	1740	1840	1940	2040	2140	2240	2340
387	433	481	532	587	645	705	770	840	910	985
532	570	610	652	695	740	787	845	885	935	987

RATING TABLES
GAGE READING VS. CFS DISCHARGE

River	Gage Location	4.6	4.7	4.8	4.9	5.0	5.1	5.2
Ammonoosuc	Bath	1550	1630	1710	1790	1870	1960	2050
	Bethlehem	1050	1110	1170	1230	1290	1360	1430
Androscoggin	Errol	4760	4915	5072	5238	5409		
Assabet	Maynard	806	857	910	965	1020	1080	1140
Ashuelot	Gilsum Gorge	340	360	381	403	425	449	473
	Hinsdale	780	862	950	1042	1140	1237	1340
S. Br. Ashuelot	Webb (Rte. 12)	340	385	435	483	535	590	643
Blackwater	Webster	590	636	684	734	786	839	892
Branch	Forestdale	708	750	792	835	880	920	962
Cold	Drewsville	446	482	519	559	600	646	694
Contoocook	Peterborough	1265	1330	1396	1464	1534	1606	1679
	Henniker					110	125	142
Deerfield	Charlemont					3150		
Ellis	Jackson (Rte. 16)							
Farmington	New Boston	673	728	786	847	910	971	1040
	Tariffville	3540	3680	3820	3960	4100	4240	4380
Green	E. Colrain	818	882	948	1020	1090		
Housatonic	Gaylordsville	2610	2730	2860	2990	3120	3250	3390
Hudson	N. Creek	2441	2585	2736	2891	3053	3320	3392
Mascoma	Mascoma (Rte. 4A)					2681		
Millers	S. Royalston	200	233	269	306	347	390	436
	Farley	1830	1930	2040	2160	2280	2410	2540
North	Shattuckville	912	988	1070	1150	1240		
Piscataquog	Grasmere	335	372	412	455	500	549	600
Quaboag	W. Brimfield	555	596	638	682	728	775	824
Salmon	Comstock Bridge	460	520	580	650	720	790	860
Sandy	Robertsville	1040	1100	1160	1220	1280	1350	1420
Saxtons	Saxtons River	340	379	420	452	490	530	572
Sacandaga	Griffin	656	694	733	774	816	859	904
	Hope	3600	3832	4072	4321	4580	4825	5078
Smith	Bristol	665	713	763	815	870	914	959
Sugar	W. Claremont	2309	2420	2533	2650	2768		
Ten Mile	Webatuck	1510	1580	1660	1740	1820		1990
West	Jamaica	137	162	189	220	255	295	332
Westfield	Knightville	737	792	850	909	970	1030	1100
	Huntington	2440	2550	2660	2770	2880	3000	3110
White	W. Hartford	1060	1140	1220	1300	1380	1430	1570
Williams	Brockway Mills	1040	1095	1149	1204	1261	1319	1379

Gage reading (feet)

5.3	5.4	5.5	5.6	5.7	5.8	5.9	6.0	6.1	6.2	6.3
2140	2230	2320	2410	2500	2600	2700	2800	2920	3040	3160
1500	1570	1650	1730	1810	1890	1980	2070	2170	2270	2380
		6307					7329			
1200	1260	1330	1400	1460	1530	1600	1680	1750	1820	1900
499	525	552	580	608	638	669	700	735	771	808
1438	1540	1649	1762	1880	2008	2141	2280	2424	2574	2730
700	760	823	889	960	1028	1098	1172			
946	1003	1062	1122	1182	1244	1308	1374			
1000	1050	1090	1130	1180	1220	1260	1310	1350	1400	1450
745	798	853	912	973	1037	1104	1174	1247	1323	1402
1754	1831	1909	1989	2071	2155	2240	2325			
160	180	200	220	245	270	300	330	360	390	430
		3960					4860			
1100	1170	1240	1310	1390	1470	1550	1630	1720	1800	1890
4520	4660	4800					5600			
3530	3670	3820	3970	4120	4280	4440	4600	4760	4930	5100
3571	3755	3945	4140	4342	4550	4750	4945	5150	5361	5576
486	540	591	646	704	765	830	899	972	1050	1130
2680	2820	2970	3110	3270	3430	3590	3750			
		1720					2320			
651	704	760	818	880	944	1010	1080	1146	1215	1286
874	926	980	1030	1090	1156	1210	1270	1330	1400	1460
930	1010	1090	1170	1250	1330	1420	1510	1600	1690	1790
1490	1560	1630	1700	1770	1840	1920	2000			
617	663	712	763	816	872	930	990	1050	1110	1180
950	996	1044	1073	1144	1196	1249	1304	1360	1418	1478
5337	5605	5880	6142	6410	6689	6983	7284	7592	7907	8230
1006	1054	1103	1154	1206	1259	1314	1370			
	2170		2350		2550		2750		2960	
373	418	467	521	580	635	695	759	827	900	977
1170	1240	1310	1390	1470	1550	1630	1710	1800	1880	1970
3230	3360	3480	3610	3740	3870	4000	4140			
1670	1770	1870	1980	2090	2200	2310	2420	2540	2660	2760
1440	1500	1562	1624	1689	1754	1821	1889	1959	2030	2098

APPENDIX VII cont'd

RATING TABLES
GAGE READING VS. CFS DISCHARGE

River	Gage Location	6.4	6.5	6.6	6.7	6.8	6.9	7.0
Ammonoosuc	Bath	3280	3400	3520	3640	3760	3880	4000
	Bethlehem	2500	2620	2740	2860	2980	3100	3220
Androscoggin	Errol		8369					9442
Assabet	Maynard	1970	2050	2130	2220	2290	2370	2450
Ashuelot	Gilsum Gorge	846	885	926	967	1010	1055	1100
	Hinsdale	2892	3060	3235	3417	3605	3799	4000
S. Br. Ashuelot	Webb (Rte. 12)		1573					2024
Blackwater	Webster		1744					2171
Branch	Forestdale	1490	1540	1590	1630	1680	1730	1780
Cold	Drewsville	1484	1570	1653	1739	1828	1920	2016
Contoocook	Peterborough							
	Henniker	470	510	550	590	630	675	725
Deerfield	Charlemont		5850					6930
Ellis	Jackson (Rte. 16)							
Farmington	New Boston	1990	2080					
	Tariffville		6400					
Green	E. Colrain							
Housatonic	Gaylordsville	5270	5450					6380
Hudson	N. Creek	5796	6021	6251	6486	6727	6973	7224
Mascoma	Mascoma (Rte. 4A)							
Millers	S. Royalston	1210	1290	1380	1470	1570	1660	1770
	Farley		4610					
North	Shattuckville		3030					
Piscataquog	Grasmere	1360	1437	1517	1600	1685	1772	1862
Quaboag	W. Brimfield	1530	1600	1670	1740	1810	1890	1960
Salmon	Comstock Bridge	1810	1990	2090	2200	2310	2420	2530
Sandy	Robertsville							
Saxtons	Saxtons River	1250	1310	1390	1460	1540	1620	1700
Sacandaga	Griffin	1539	1602	1666	1732	1799	1868	1939
	Hope	8570	8918	9274	9638	10,010	10,390	10,780
Smith	Bristol							
Sugar	W. Claremont							
Ten Mile	Webatuck	3180		3400		3620		3860
West	Jamaica	1061	1150	1233	1320	1412	1509	1611
Westfield	Knightville	2070	2160	2260	2350	2450	2560	2660
	Huntington		4850					5620
White	W. Hartford	2910	3040	3170	3310	3450	3590	3740
Williams	Brockway Mills	2168	2239					

Gage reading (feet)

7.1	7.2	7.3	7.4	7.5	7.6	7.7	7.8	7.9	8.0	8.1
4140	4280	4420	4560	4700	4840	4980	5120	5260	5400	5560
3340	3460	3590	3720	3850	3980	4110	4240	4370	4500	
				10,540						
2530	2620	2700	2780	2870	2960	3050	3140	3230	3320	
1151	1203	1257	1312	1369	1428	1488	1550	1614	1680	1748
				5060					6300	
				2648					3192	
1830	1880	1930	1990	2040	2100	2160	2210	2270	2330	
2114	2215	2320	2428	2548	2649	2761	2876	2994	3115	3240
775	825	880	940	1000	1060	1120	1190	1260	1330	1400
				8070					9290	
				7380					8450	
7480	7742	8009	8281	8559	8842	9131	9425	9725	10,030	10,340
1870	1990	2100	2220	2350	2480	2610	2750	2900	3050	
1955	2050	2144	2240	2339	2440	2544	2650	2756	2865	
2040	2120	2200	2280	2370						
2650	2770	2900	3030	3160	3300	3440	3580	3720	3870	
				2050					2430	
				2322					2841	
				12,850					15,780	
1717	1830	1947	2071	2200					2855	
2760	2860	2970	3070	3180	3290	3400	3510	3620	3740	
				6440						
3890	4040	4190	4350	4510	4670	4840	5010	5180	5360	

APPENDIX VII cont'd

RATING TABLES
GAGE READING VS. CFS DISCHARGE

River	Gage Location	8.2	8.3	8.4	8.5	8.6	8.7	8.8
Ammonoosuc	Bath	5720	5880	6040	6200	6360	6520	6680
	Bethlehem				5150			
Androscoggin	Errol							
Assabet	Maynard							
Ashuelot	Gilsum Gorge	1818	1890	1964	2040	2118	2198	2280
	Hinsdale				7675			
S. Br. Ashuelot	Webb (Rte. 12)							
Blackwater	Webster							
Branch	Forestdale				2650			
Cold	Drewsville	3339	3440	3542	3647	3752	3860	3951
Contoocook	Peterborough							
	Henniker	1480	1560	1640	1730	1830	1930	2040
Deerfield	Charlemont				10,600			
Ellis	Jackson (Rte. 16)							
Farmington	New Boston							
	Tariffville							
Green	E. Colrain							
Housatonic	Gaylordsville				9580			
Hudson	N. Creek	10,660	10,980	11,310	11,640	11,980	12,330	12,680
Mascoma	Mascoma (Rte. 4A)							
Millers	S. Royalston							
	Farley							
North	Shattuckville							
Piscataquog	Grasmere				3453			
Quaboag	W. Brimfield							
Salmon	Comstock Bridge							
Sandy	Robertsville							
Saxtons	Saxtons River							
Sacandaga	Griffin				3208			
	Hope				17,800			
Smith	Bristol							
Sugar	W. Claremont							
Ten Mile	Webatuck							
West	Jamaica				3644			
Westfield	Knightville							
	Huntington							
White	W. Hartford				6280			
Williams	Brockway Mills							

Gage reading (feet)

8.9	9.0	9.1	9.2	9.3	9.4	9.5	9.6	9.7	9.8	9.9	10.0
6840	7000	7180	7360	7540	7720	7900	8080	8260	8440	8620	8800
	5830					6550					7310
2364											
8900											
	2990										
						3370					
2150	2270	2400	2530	2670	2810	2960	3120	3280	3450	3620	3800
	10,800										
13,040	13,400					15,300					
	4110										
	3706										
	20,690										
	4547					5576					
	7280					8380					

When you are tired of your old boat or when you have had a bad day, this is how to get rid of your boat — Dave Cooper